Formal Epistemology and Cartesian Skepticism

"This groundbreaking book is a bold, and much needed, attempt to bridge formal and traditional epistemology. It employs Bayesian reasoning to confront Cartesian skepticism and other classical philosophical puzzles. Indeed, the book even goes beyond Bayesianism and covers recent proposals on which active research is taking place at the moment."

—*Gustavo Cevolani, IMT School for Advanced Studies Lucca, Italy*

This book develops new techniques in formal epistemology and applies them to the challenge of Cartesian skepticism. It introduces two formats of epistemic evaluation that should be of interest to epistemologists and philosophers of science: the dual-component format, which evaluates a statement on the basis of its safety and informativeness, and the relative-divergence format, which evaluates a probabilistic model on the basis of its complexity and goodness of fit with data. Tomoji Shogenji shows that the former lends support to Cartesian skepticism, but the latter allows us to defeat Cartesian skepticism. Along the way, Shogenji addresses a number of related issues in epistemology and philosophy of science, including epistemic circularity, epistemic closure, and inductive skepticism.

Tomoji Shogenji teaches philosophy at Rhode Island College. His main area of research is formal epistemology, and his publications include "Is coherence truth conducive?" (Analysis 1999) and "The Degree of epistemic justification and the conjunction fallacy" (Synthese 2012) among many others.

Routledge Studies in Contemporary Philosophy

94 **Isn't That Clever**
A Philosophical Account of Humor and Comedy
Steven Gimbel

95 **Trust in the World**
A Philosophy of Film
Josef Früchtl

96 **Taking the Measure of Autonomy**
A Four-Dimensional Theory of Self-Governance
Suzy Killmister

97 **The Legacy of Kant in Sellars and Meillassoux**
Analytic and Continental Kantianism
Edited by Fabio Gironi

98 **Subjectivity and the Political**
Contemporary Perspectives
Edited by Gavin Rae and Emma Ingala

99 **Aspect Perception after Wittgenstein**
Seeing-As and Novelty
Edited by Michael Beaney, Brendan Harrington and Dominic Shaw

100 **Nature and Normativity**
Biology, Teleology, and Meaning
Mark Okrent

101 **Formal Epistemology and Cartesian Skepticism**
In Defense of Belief in the Natural World
Tomoji Shogenji

Formal Epistemology and Cartesian Skepticism

In Defense of Belief in the Natural World

Tomoji Shogenji

Routledge
Taylor & Francis Group
NEW YORK AND LONDON

First published 2018
by Routledge
711 Third Avenue, New York, NY 10017

and by Routledge
2 Park Square, Milton Park, Abingdon, Oxon OX14 4RN

Routledge is an imprint of the Taylor & Francis Group, an informa business

Library of Congress Cataloging-in-Publication Data
A catalog record for this book has been requested

ISBN: 978-1-138-57018-4 (hbk)
ISBN: 978-0-203-70377-9 (ebk)

Typeset in Sabon
by Apex CoVantage, LLC

For Yao

Contents

Acknowledgments

Most of the materials presented in this book have not appeared before, but portions of it draw on my previous work. Part of Chapter 1 (Sections 4–5) is based on “Internalism and externalism in meliorative epistemology”, *Erkenntnis* 76 (2012), 59–72. Part of Chapter 3 (Sections 1–5) is an extension of a series of articles on epistemic circularity: “Self-dependent justification without circularity”, *The British Journal for the Philosophy of Science* 51 (2000), 287–298; “A defense of reductionism about testimonial justification of beliefs”, *Noûs* 40, 331–346; and “Reductio, coherence, and the myth of epistemic circularity” in F. Zenker, ed. *Bayesian Argumentation* (Springer 2013), 165–184. Part of Chapter 4 (Sections 2–4 and Appendices) is based on “The degree of epistemic justification and the conjunction fallacy”, *Synthese* 184 (2012), 29–48. Part of Chapter 5 (Sections 1–2) draws on the joint article with Luca Moretti, “Skepticism and epistemic closure: Two Bayesian accounts”, *International Journal for the Study of Skepticism* 7 (2017), 1–25.

Some colleagues read and commented on part or whole of the manuscript at various stages of its development. I would like to thank Harold Brown, Matt Duncan, Chang-Seong Hong and especially William Roche, for many thoughtful comments. I would also like to thank anonymous reviewers of *Routledge* for valuable suggestions on the penultimate version of the manuscript.

Introduction

Cartesian skepticism questions our belief in the natural world. We hold numerous beliefs about the natural world; some of them are perceptual beliefs based on our sensory experience while others are testimonial beliefs based on reports from others. Those beliefs may be preserved in memory, and then combined and extended by deductive and inductive reasoning. Of these types of beliefs, the target of the Cartesian skeptic is perceptual beliefs because they are the most fundamental about the natural world. Testimonial beliefs are dependent on sensory experience in two ways. We receive testimony through our own sensory experience either by hearing it or by reading it. Further, if a report on the natural world is to be trusted, someone must have had sensory experience that supports its content. The report may be preserved in memory and passed on through a chain of communication, but someone at the origin of the chain must have had appropriate sensory experience at some point. Similarly, if my memory of the natural world is to be trusted, I must have had sensory experience in the past that supports its content. Inference may extend our beliefs about the natural world, but they are not the original source. So, if our perceptual beliefs cannot to be trusted, then our beliefs about the natural world are in serious doubt.

The Cartesian skeptic issues a challenge in the form of an alternative account of our sensory experience. Descartes considered an evil demon who produces misleading sensory experience in our mind. The standard scenario in contemporary epistemology is the brain-in-a-vat story where a human brain is kept alive in a vat and wired to a supercomputer programed to send misleading sensory signals to deceive the brain. The point of these scenarios is that the victims of the deception have the same kind of sensory experience they would have in the natural environment. So, if we are the victims of the deception in the scenario, we do not notice our predicament. The Cartesian skeptic argues that for all our confidence, there may not be the natural world we think we live in. The question for the epistemologist is whether there is some way of refuting these alternative accounts in defense of belief in the natural world.

No serious attempts have been made in recent epistemology to refute the alternative accounts with absolute certainty. Contemporary epistemologists are fallibilists with regard to beliefs about the natural world. It is not necessary, by their less ambitious standards, to refute the alternative accounts with absolute certainty so long as we have sufficiently strong reason for rejecting them. But what reason do we have in support of belief in the natural world if our sensory experience is indistinguishable under the alternative scenario? I see a hand in front of me—or so I think—but I will have the same sensory experience if I am a brain in a vat manipulated by a well-programed supercomputer. The scenario may seem far-fetched and improbable, but there is no obvious way of demonstrating that it is improbable.[1] There is indeed no consensus on how best to respond to the challenge of Cartesian skepticism. My project in this book is not reviewing numerous proposals and criticisms.[2] Instead, the book takes a new approach with two distinctive features.

First, the book makes full use of newly emerging concepts and principles of formal epistemology. Formal epistemologists turn to results and techniques of relevant fields in mathematics to clarify and solve issues in epistemology. It has seen much progress in recent years with significant applications; in particular, Bayesian epistemology that makes use of the formal theory of probability has been fruitful. However, there has been no sustained work in formal epistemology to respond to Cartesian skepticism. The book fills this lacuna. The second distinctive feature of the book is the framework of meliorative epistemology (Chapter 1). It is a broadly Cartesian framework in that epistemic evaluation is to guide our epistemic practice, instead of accommodating and systematizing our pre-theoretical judgments, as is common in recent epistemology. Meliorative epistemology turns to intuition only in the context of discovery where various proposals are formulated, while the basis of their evaluation is the alethic goals of cognition and the constraints of epistemic resources available to the community.[3] This means that the book takes Cartesian skepticism seriously: If we cannot meet the challenge of Cartesian skepticism with the available resources, then meliorative epistemology recommends that we concede defeat and set a less ambitious goal for our epistemic practice.

In regard to substance, the book defends two major theses. The first is that Bayesian epistemology not only fails to support our belief in the natural world, but endorses Cartesian skepticism. The reasoning for this thesis proceeds in two stages. In the first stage (Chapter 4) I argue for the dual-component format of epistemic evaluation, which is to replace the "Lockean" format of epistemic evaluation that is common in Bayesian epistemology. According to the Lockean format, there is justification for accepting the hypothesis h just in case its probability is sufficiently high, or $P(h \mid e \wedge b) \geq t$ to put it formally, where e is the body of evidence, b is the body of background beliefs, and t is the threshold of sufficiency. I show by a variant of the lottery paradox that the Lockean format allows the "backdoor endorsement" of an unacceptable hypothesis: The subject can accept (by the Lockean format) each of the probabilistically independent statements $p_1, \ldots, p_n$ to derive their logical consequence h even if h is not acceptable (by the Lockean format). The problem arises because the Lockean format only takes into account the safety of accepting the hypothesis. I propose the dual-component format of epistemic evaluation that takes into account not only the safety of accepting the hypothesis but also the informativeness of the hypothesis. I introduce the specific measure J of epistemic worthiness for the dual-component format, and show that it is the only measure (up to ordinal equivalence) that satisfies the formal constraints necessary for blocking backdoor endorsement.

I then proceed to the second stage (Chapter 5) where Cartesian skepticism is vindicated by the dual-component format of epistemic evaluation. This is not because the natural world hypothesis has a low degree of epistemic worthiness. Instead the natural world hypothesis is superseded by the disjunction of the natural world hypothesis and the skeptic's alternative hypothesis in the sense that the disjunction has a higher degree of epistemic worthiness and once the disjunction is incorporated into the subject's body of beliefs, the natural world hypothesis loses its epistemic worthiness relative to the expanded body of beliefs. Supersedure by the disjunction lends support to the skeptic's position that we should withhold judgment between the natural world hypothesis and the alternative hypothesis. I show more generally that the disjunction of empirically equivalent hypotheses supersedes each disjunct, so that Bayesians should aim at epistemic justification only up to empirical equivalence.

The second major thesis of the book is the rejection of Cartesian skepticism, despite its Bayesian support, based on the evaluation of probabilistic models. The reasoning for this thesis also proceeds in two stages. In the first stage (Chapter 6) I introduce the relative-divergence format of epistemic evaluation for probabilistic models, which is to supplement the dual-component format that makes use of probabilities. A model is only a template of theories and generates a theory from the given data by filling in parameter values that best fit the data. It is known in statistical learning theory that although a complex model with more adjustable parameters can fit the data better, it does not always capture the true relation better because the flexibility of a complex model makes it more prone to chase random variation ("noise") in the data. The relative-divergence format therefore evaluates probabilistic models by balancing the complexity of the model and the model-generated theory's goodness of fit with the data.

In the second stage (Chapter 7) I argue that the natural world model is superior to the skeptic's alternative model by the relative-divergence format of epistemic evaluation. The reason is that the skeptic's model consists of two parts—the model of the virtual world that matches the natural world model, and the model of the causal process by which the deceiver produces sensory experience in our mind. In contrast, the natural world model already contains the causal process by which the natural environment produces sensory experience in our mind because the process is part of the natural world. The added complexity due to the process part of the skeptic's alternative model makes it inferior to the natural world model by the relative-divergence format of epistemic evaluation. Cartesian skepticism is thereby defeated.

The book is organized around Cartesian skepticism, but it addresses a number of related issues along the way, including the conception of truth (Chapter 2), epistemic circularity (Chapter 3), epistemic closure (Section 1 of Chapter 5) and inductive skepticism (Section 6 of Chapters 5; Sections 1–2 of Chapter 7). These discussions should be of interest to epistemologists and philosophers of science beyond the context of Cartesian epistemology. The two formats of epistemic evaluation presented in the book are also applicable to issues of epistemology and philosophy of science in general. These are dividends of taking the skeptic's challenge seriously and confronting it with all relevant resources available.

Notes

1 Note that the laws of nature familiar to us may not hold in the world we actually belong to according to the alternative scenario. We cannot simply cite these laws to assign a low probability to the alternative scenario.
2 See Lyon (2016) for an overview of major proposals and criticisms.
3 Meliorative epistemologists may look into our pre-theoretical judgments *after* they resolve an issue in epistemology to see whether our current epistemic practice—exemplified by our pre-theoretical judgments—is in need of revision.

1 Skepticism and the Method of Meliorative Epistemology

1.1 Meliorative Epistemology

This book examines the epistemic status of our belief in the natural world. The aim is to evaluate our basic view of the natural world; for example, that familiar objects such as rocks, trees and chairs exist in the natural world where we also reside, and we interact with these objects.[1] I am going to examine whether our basic view of the natural world withstands the challenge of Cartesian skepticism. The point of dispute is not occasional errors, but the possibility of a wholesale misconception of reality. The Cartesian skeptic questions our basic view of the natural world by introducing radical alternative scenarios, e.g. that all objects of our sense perception are illusions created by a supernatural being, or that they are entities in the virtual world to which we do not belong. So, to put our basic view of the natural world in negative terms, the objects of our sense perception are not illusions created by a supernatural being, not entities in the virtual world to which we do not belong, etc. I call it "the natural world hypothesis".

In its original challenge (Descartes 1984) the skeptic introduces the possibility of a deceiver with an extraordinary power—we are actually bodiless souls but are deceived by a powerful demon into thinking that we live in a natural world.[2] For the purpose of exposition, however, I will use a popular variant discussed widely in contemporary philosophy that we are actually brains in vats (BIVs) kept alive and wired to a supercomputer that is programed to supply us with the kind of sensory input brains in skulls would receive from the natural environment. The BIV version does not deny the existence of a natural world—there are brains, vats, a supercomputer, etc.—but it denies that familiar natural objects exist in places where our sense perception locates them relative to us. What we take to be natural objects belong to the virtual world to which we do not belong.

A remark is in order on the extent of skepticism addressed in this book, especially on its impact on the scientific understanding of the natural world. Though Cartesian skepticism questions our basic view of the natural world, it does not bring back the discredited hypotheses of the past, such as the phlogiston theory and the geocentric theory. The Cartesian skeptic does not question the substance of currently accepted scientific theories, but only their interpretation. For example, the heliocentric theory may not be the correct theory of the natural world, but the correct theory of the virtual world to which we do not belong.

As stated already, I take this challenge seriously, and consider it an open question whether there is epistemic justification for accepting the natural word hypothesis. It is sometimes suggested that the cost of giving up the natural world hypothesis is too high, so that it is sensible to reject some element of the skeptic's argument even if the argument appears compelling. It is not clear, however, what the high cost amounts to. There is a sense in which whether we accept the natural world hypothesis or the BIV hypothesis makes no difference to us because the two hypotheses are empirically equivalent—we will enjoy (or suffer from) the same sensory experience if we are BIVs instead of living in a natural environment. What is the cost there? Some people may mention the loss of familiarity and intuitiveness. Since we commonly take the natural world hypothesis for granted, any view that is inconsistent with it, such as the BIV hypothesis, is counterintuitive. But if the cost is the loss of familiarity and intuitiveness, its relevance to epistemic evaluation is questionable. Why are familiarity and intuitiveness important in epistemic evaluation?

Over the course of history, numerous theories that used to be familiar and intuitive were replaced by unfamiliar and counterintuitive theories based on new evidence and new conceptual tools. It may still be counterintuitive to some people that the earth is orbiting around the sun instead of the sun orbiting the earth, but the educated public have come to terms with the heliocentric theory. It is possible in a similar way that we may eventually come to terms with a counterintuitive theory with regard to the natural world as a whole. Indeed, our stance on the natural world hypothesis has already changed due in part to Cartesian skepticism. People with no exposure to philosophical reflection may still believe it is absolutely certain that the natural world hypothesis is correct. There is, however, a strong sense among epistemologists that we cannot rule out the BIV hypothesis with absolute certainty. This does not immediately lead to a surrender to skepticism. Most epistemologists no longer share the Cartesian aspiration for absolute certainty. Instead, they accept fallibilism and are content with the probabilistic evaluation of the hypothesis.

Given the precedence of the retreat to fallibilism, it is conceivable that the challenge of Cartesian skepticism pushes us farther away from the familiar and intuitive view to the reluctant concession that the natural world hypothesis is indefensible even probabilistically. As we will see in Chapter 5, a strong Bayesian case can be made for accepting Cartesian skepticism and retreating to the goal of epistemic evaluation up to empirical equivalence. In the investigation that follows, the import of philosophical skepticism is its possible role in the revision of our epistemic practice, including our basic view of reality and the prevailing format of epistemic evaluation. The pretheoretical judgment to the contrary is not good reason to resist revision.

The basic question of methodology arises at this point: What is the basis of revising or reaffirming our epistemic practice if we cannot turn to our pre-theoretical judgments? My answer is meliorative epistemology, which is broadly a Cartesian approach to epistemology.[3] Cartesian epistemology is not grounded in our pre-theoretical judgments, but is meant to guide them. As Descartes put it, "The aim of our studies should be to direct the mind with a view to forming true and sound judgments about whatever comes before it" (Descartes 1985a, p. 9). More generally, in meliorative epistemology epistemic evaluation is expected to guide our epistemic practice to better results, where the better results are understood alethically. When our cognitive state is qualitative—to accept a statement, reject it, or withhold judgement—the alethic goal is to increase true beliefs and avoid false beliefs. When our cognitive state is quantitative—to assign a probability to a statement—the alethic goal is to make the probability as close to the true value as possible.[4] This is all done under the resource constraints, i.e. we pursue the alethic goals with the epistemic resources available to us, both empirical and conceptual. The format of epistemic evaluation should be determined so that it will best serve the alethic goals under the resource constraints.

The project of meliorative epistemology is similar to a project in engineering. Engineers seek the best design for achieving certain goals under the given constraints. Similarly, meliorative epistemologists seek the best format of epistemic evaluation to achieve the alethic goals under the constraints of epistemic resources available. We may call it conceptual engineering.[5] As in any engineering project, the constraints can change over time. As new empirical evidence emerges or new conceptual tools are devised, the best format of epistemic evaluation may also change. The alethic goals are not set in stone either. We may aim at a more ambitious goal as the epistemic resources expand, but it is also possible that we need to give up the original goal in favor of a less ambitious goal if new evidence or new conceptual tools reveal that the original goal is unachievable. That is one of the open questions addressed in this book.

1.2 The Role of Intuition

I adhere to the policy of not relying on our pre-theoretical judgments in adjudicating competing theories of meliorative epistemology. Since the policy is not widely shared among contemporary epistemologists, I want to address likely objections to the policy in this section.[6] Let us see first how an argument from pre-theoretical judgment typically works. In his celebrated article (Gettier 1963), Gettier challenged the traditional theory of knowledge, according to which *S* knows that *p* if and only if *S* has a justified true belief that *p*. Gettier used two imaginary cases (thought experiments) to undermine the theory. In one of them (the ten coins case) the description of the circumstance makes it clear that Smith—an applicant for a certain job—has a justified true belief that the man who will get the job has ten coins in his pocket, but the description also makes it clear that Smith does not know this proposition.[7] The traditional theory of knowledge is thereby rejected. The key in this argument is the judgment that Smith does not know the proposition in question, and this judgment is not formed by conscious inference from the information given. The argument is supported by a pre-theoretical judgment, or "intuition" as it is called in philosophical discussions. The use of intuition in thought experiments is not confined to epistemology. The trolley case in ethics (Foot 1967) and the Gödel-Schmidt case in philosophy of language (Kripke 1980) are often mentioned in the discussion of intuition in philosophy.

It is usually conceded by its advocates that intuition as evidence is defeasible. It can be overridden by other intuitions, other types of evidence, or by some theoretical considerations. According to this view, pre-theoretical judgments are not non-negotiable data for the analysis. However, some people may think I have gone too far in rejecting any use of intuition in meliorative epistemology. Indeed it is doubtful that we can avoid the use of pre-theoretical judgments completely. How can we start any epistemological investigation, or any investigation for that matter, without relying on some pre-theoretical judgments? Even scientists in search of an explanation or engineers seeking the best design routinely rely on their intuitive judgments, and no logical inference from consciously identified information can replace them because intuitive judgments have different roots, most likely in the subpersonal network of associations.[8] Of course, the kind of intuitive judgments that scientists and engineers make are informed by their rich past experience as researchers and are often abstract in nature, while intuitive judgments in typical thought experiments are non-technical and concrete. But they are still of the same kind in the sense that they are, presumably, both rooted in the subpersonal network of associations, and not derived by explicit inference from consciously identified information.

It may be thought that I am pragmatically incoherent because I must have relied on my intuition in some way in the process of constructing an argument against the use of intuition in meliorative epistemology. We may call this line of argument "the self-defeat argument".[9] In response to the self-defeat argument, I want to call attention to the distinction between the context of discovery and the context of justification.[10] I grant that we need intuition to conduct any investigation. Scientists in search of an explanation often rely on their intuition in generating new hypotheses or recognizing their potential shortcomings. Engineers seeking the best design also rely on their intuition to come up with new ideas or anticipate problems to overcome. Investigators in any field consult their intuition, among many other factors, to decide on the direction of their research. However, this is all in the context of discovery. Scientists and engineers do not (and should not) cite their intuition in the context of justification, e.g. when the scientists defend their theory in a scientific article or when the engineers explain the merit of their design to their client. It is inappropriate to say "I have the intuition that this is correct" in the context of justification.

The same is true in meliorative epistemology. In order to generate hypotheses about good epistemic practice or to detect their potential problems, meliorative epistemologists may rely on their intuition. They may also consult their intuition, among many other factors, to decide on the direction of their research. However, this is all in the context of discovery. It is as inappropriate for them to cite their intuition in the context of justification as it is for the scientists or the engineers to do so. This is also true of objections to some theory. The opponents of a theory can certainly rely on their intuition in the context of discovery, e.g. in their search for points against the proposed theory. However, it is inappropriate for them to cite their intuition as the ground of their objection.[11] The arguments in meliorative epistemology, both for and against a theory, must be based on the alethic goals of our epistemic practice and the constraints of our epistemic resources.

Some defenders of intuition may point out that we would have no good reason to use intuition even in the context of discovery unless it is genuine evidence, though perhaps weak and easily defeasible evidence. Why should we use it in generating hypotheses and deciding on the direction of research if intuition were not even weak and easily defeasible evidence?[12] Though scientists and engineers avoid citing intuition in their formal arguments and presentations, its weakness as evidence and easy defeasibility explain their reluctance, viz. formal arguments and presentations call for evidence that is strong and not easily defeasible. It does not mean that intuition has no evidential value at all. In philosophy, where strong evidence is hard to come by and we do not expect to remove all reasonable doubt, intuition is admissible even in the context of justification, it may be suggested.

I am willing to grant that intuition is not totally unreliable in the sense of being a random occurrence. However, it is impossible to measure the reliability of intuition in a meaningful way.[13] The reliability of intuition depends on the amount and the quality of the relevant experience the particular individual has on the particular subject. We cannot extrapolate from the past successful or unsuccessful use of intuition by other people on different subjects. There may be some philosophers who think that their discipline is in such a desperate state that even evidence from a source to which no meaningful degree of reliability can be assigned is admissible in the context of justification. I disagree, but they are free to pursue their project as they deem fit provided they make it clear that their evidence comes from a shaky source, and their conclusion is speculative. In this book I will consult intuition only in the context of discovery, and discuss it only when I introduce an idea informally and explain the motivation. In the context of justification, I will disregard intuition whether it is for or against the proposed theory. I will often say that intuition plays no role in meliorative epistemology, ignoring its role in the context of discovery, because it is true of any intellectual project that intuition plays a role in the context of discovery.

1.3 Meliorative Epistemology vs. Epistemography

Intuition plays no role in the evaluation of a theory in this book, but this is not a polemical stance I take in the sense that I reject other approaches in epistemology. It may shed light on our epistemic practice to formulate general principles that govern our pre-theoretical judgments of the epistemic kind. If that is the project, which I call *epistemography*, then the formulated general principles should accommodate our pre-theoretical judgments. It is therefore perfectly fine for an epistemographer to cite our pre-theoretical judgments in adjudicating competing hypotheses. There may also be other valuable projects in epistemology such as deontological epistemology to determine our obligation as responsible epistemic subjects. I have no objection to any such projects, provided the advocates of each project are clear and upfront about the nature of their project—what they try to accomplish and what are the grounds for adjudicating competing hypotheses.

The worst approach is to mix together projects with different goals and grounds. For example, we should not adjust a format of epistemic evaluation in meliorative epistemology to accommodate our pre-theoretical judgments; we should not disregard pre-theoretical judgments in epistemography to make them consistent with some principles of epistemic evaluation. Some epistemologists defend adjustments of this kind in the name of *reflective equilibrium*, but it is unclear what project in epistemology is served by the method. When successfully executed, the method of reflective equilibrium may produce a coherent set of principles that accommodates most of our pre-theoretical judgments. The result is often appealing to many people. Who would dislike a coherent system that is mostly in accord with their intuition? However, the question is what it accomplishes theoretically. Neither meliorative epistemology nor epistemography is well served by the method of reflective equilibrium.

As explained in the previous section, I have no objection to consulting intuition in the context of discovery. The method of reflective equilibrium may be understood charitably as a way of formulating a hypothesis in the context of discovery. That is perfectly fine. I am not ruling out the possibility, either, that some legitimate project in epistemology may be served well by the method of reflective equilibrium. In fact it is not difficult to see the value of a reflective equilibrium in certain fields of applied philosophy where building a consensus is important. For example, in normative political theory, to which Rawls (1971) introduced the notion of a reflective equilibrium, it is crucial to build a consensus among members of the community in order to secure the sense of legitimacy.[14] The method of reflective equilibrium is a sensible approach when building a consensus is important, but it is hard to find good reason for adopting the method of reflective equilibrium in epistemology. There is certainly no good reason for adopting it in meliorative epistemology.

Going back to the project of epistemography, it is legitimate to cite our pre-theoretical judgments for or against a hypothesis in epistemography. However, since they are systematizing pre-theoretical judgments of some community, epistemography cannot resolve disputes between different communities (sometimes between different sub-communities, and even different individuals).[15] For example, those who are committed to certain beliefs—perhaps religious or moral—may come up with some system of epistemic principles that bestows those beliefs a high epistemic status, but their opponents who question the beliefs will also question the principles. Since the two communities start from different sets of pre-theoretical judgments, they cannot resolve the dispute over the principles. To put this in a cynical way, epistemography only systematizes the epistemic prejudice of some community. This is true even if there is no serious disagreement in pre-theoretical judgments among different communities. Epistemography only systematizes their shared epistemic prejudice.

It may seem that at least epistemography can remove ambiguity in the pre-theoretical judgments, but that is not true either. Suppose most people in the community are unsure of the epistemic status of some beliefs, e.g. whether or not the epistemic subject is justified in holding a certain belief under some unusual conditions. It may seem that we can remove ambiguity by formulating general principles on the basis of unambiguous cases, and then apply the formulated principles to obtain a clear-cut answer in the ambiguous case. However, if we are searching for general principles that govern our pre-theoretical judgments and our pre-theoretical judgment is ambiguous in the case, then the general principles should reflect the ambiguity of our pre-theoretical judgment. In other words, we should question general principles that give clear-cut answers in cases where our pre-theoretical judgment is ambiguous.

This is not to say that epistemography should allow no discrepancy between general principles and our pre-theoretical judgments. Since there may be non-epistemic "noise" in our seemingly epistemic judgments, it is possible that highly complex principles that accommodate all our pre-theoretical judgments are actually chasing some non-epistemic noise instead of capturing the principles that govern our epistemic judgments. The epistemographers may therefore prefer simple principles with minor discrepancies with our pre-theoretical judgments to highly complex principles with no discrepancies. The point mentioned earlier, however, is that in cases where our pre-theoretical judgment is ambiguous, "no discrepancy" means that the general principles reflect the ambiguity of our pre-theoretical judgment. There is a discrepancy if our pre-theoretical judgment is ambiguous while the general principles give a clear-cut answer.

Despite all these caveats, epistemography can still be part of a broader research program. For example, it is sensible to make the presumption that conceptual tools we have been using for many centuries have some useful functions, and this presumption points to the possibility of not just uncovering the principles that govern our pre-theoretical judgments, but identifying their functions.[16] If meliorative epistemology is a project in conceptual engineering, where we seek the best format of epistemic evaluation for achieving the alethic goals under the resource constraints, epistemography can be considered part of reverse conceptual engineering, where we search for the function of the principles that govern our pre-theoretical judgments. It is comparable to an engineer who comes across a mysterious device in a rival company's new product and tries to figure out its function from the way it is composed, or to a biologist who comes across an intricate structure in an organism and tries to figure out its function from the way it is composed.

Of course, the presumption does not justify the use of pre-theoretical judgments in meliorative epistemology. Even if an apparently epistemic concept, such as knowledge, serves some function well, a significant part of their function may turn out to be social or even political, and unrelated to the alethic goals of increasing true beliefs and avoiding false beliefs or assigning a probability as close to the true value as possible. More importantly, as greater epistemic resources become available, the best design for achieving the alethic goals can also change.[17] The Cartesian skeptic introduces a novel conceptual tool in the form of the BIV hypothesis that is empirically equivalent to the natural world hypothesis. This was not part of the epistemic resources available to our ancestors. The pre-theoretical judgments we inherited from our ancestors may no longer serve the alethic goals well in light of the new conceptual tool. Note in particular that we cannot argue against Cartesian skepticism on evolutionary grounds that we would not have survived if our view of the natural world were radically mistaken. There is no problem of survival if the BIV hypothesis is correct because it is empirically equivalent to the natural world hypothesis—our virtual bodies are well adjusted to survive in the virtual world generated by the supercomputer.

1.4 Resource Constraints

This section examines the basis of epistemic evaluation in meliorative epistemology. As explained already, if epistemic evaluation is to guide our epistemic practice, the basis of the evaluation must be the epistemic resources available to us. However, we can interpret this constraint in two ways. The basis may be either the resources available to the evaluator or the resources available to the subject whose belief is evaluated. If it is the latter (Nagel 1989, Ch. 5; Pollock and Cruz 1999, Ch. 1; Stevenson 1999), then meliorative epistemology inherits the traditional internalist constraints on epistemic evaluation. However, some recent meliorative epistemologists (Kitcher 1992) drop the traditional constraints in favor of less restrictive constraints. Some forms of externalism are clearly unsuitable. For example, we cannot say in meliorative epistemology that beliefs are in good epistemic standing provided they are actually produced by a reliable process (Goldman 1979), or provided they are actually formed by a properly functioning apparatus (Plantinga 1993), even if no one can tell whether the process is reliable, or the apparatus is functioning properly. This is because we cannot improve our epistemic performance unless we evaluate beliefs on the basis of the epistemic resources available to us. Instead, the idea suggested is to drop the traditional internalist constraints and let the better-informed *experts* among us to evaluate the beliefs based on the resources available to them.

The division of cognitive labor is a powerful tool for improving our epistemic practice. Though each member of the community has limited epistemic resources on most subject matters, they can turn to experts in the relevant field in their community, who can draw on much better resources. Social aspects of our epistemic practice have been emphasized by an increasing number of epistemologists (e.g. Schmitt 1994; Goldman 1999). It seems absurd to enforce the internalist constraints and disregard the wealth of information obtained in all kinds of scientific research, including research on the human cognitive process.[18] We can benefit from discoveries on the formation of beliefs, just as we do from discoveries on the contents of beliefs, such as climatology and medicine. Meliorative externalists therefore urge us to take full advantage of naturalized and social epistemology, in which the expert's evaluation is based on evidence and reasoning unavailable to the ordinary epistemic subject. This is not an abstract possibility. Take, for example, the research program of "heuristics and biases" in psychology (Tversky and Kahneman 1974) and its application in behavioral economics (e.g. Ariely 2008; Thaler and Sunstein 2008). Works in this research program have uncovered surprising weaknesses in our cognitive process that have clear implications in meliorative social epistemology, and are beginning to influence our epistemic practice. In general it is sensible to let the experts evaluate the formation of beliefs and learn from it. Of course, the epistemic experts cannot evaluate each community member's beliefs individually, but they can provide epistemic guidelines to follow.[19]

Epistemic guidelines expand the range of beliefs that benefit from the division of cognitive labor. Although there are experts on the subjects of major significance such as climatology and medicine, there are usually no experts who can evaluate the contents of our everyday beliefs on personal matters. For example, there is no expert who can tell the location of your misplaced keys. However, you may still get some help from a general guideline on memory. Any belief we form can be the subject of some epistemic guideline. Since an epistemic guideline is only a general instruction, each epistemic subject must identify certain conditions in specific cases they face. For example, if memory-based belief is reliable under certain conditions according to some epistemic guideline, then the epistemic subject must figure out whether these conditions hold in cases of her interest. This means that the resource constraints in meliorative epistemology are those of "internalist externalism" (Alston 1988) in regard to our everyday beliefs on personal matters: The epistemic subject needs to have an access to the ground for holding her belief, but she need not have sufficient evidence for the adequacy of the ground. It is the epistemic expert's responsibility to evaluate the adequacy of the ground in the form of an epistemic guideline. Of course, when the expert in some field directly evaluates the content of the belief, the epistemic subject who holds the belief need not have an access to the ground for her belief.

Despite the obvious benefit from the division of cognitive labor, there is also good reason to take the traditional internalist constraints seriously. A well-informed epistemic evaluation by the experts will not improve the subject's epistemic performance unless she accepts the evaluation in light of the evidence and reasoning available to her. For example, when she receives an epistemic guideline according to which her latest belief is poorly supported by the evidence, she must judge whether the guideline is reliable. There are also many cases where no expert counsel is available, and thus the epistemic subject has no choice but to evaluate the statement for herself. Either way, each epistemic subject must rely on their own epistemic resources in the end in to improve their epistemic performance.

I just sketched two ways of understanding the resource constraints in meliorative epistemology. The meliorative externalist stresses the benefit from the division of cognitive labor, while the meliorative internalist points out the need for the subject's own evaluation of the expert counsel. Both sides have good points and we need not choose one side at the expense of the other because epistemic evaluation in meliorative epistemology comprises two components. At the social level—meliorative social epistemology—appropriate experts should evaluate particular statements and general guidelines (hereafter both of them will be called "statements") for dissemination in the community, while at the personal level—meliorative personal epistemology—each member of the community needs to evaluate the expert counsel for their personal use. Epistemic evaluation at the social level is externalist in the sense that the basis of evaluation need not be accessible to each member of the community, while epistemic evaluation at the personal level is internalist.

It should be noted that the personal evaluation of the expert's counsel is indirect in most cases. For example, most epistemic subjects can only judge—if at all—whether those who endorsed statements are competent and honest, instead of replicating their research for themselves by spending many hours in the lab. If the epistemic subject judges that they are competent and honest, then she has good reason to accept their counsel. Indirect evaluation is a departure from the traditional form of internalism, where each epistemic subject is expected to evaluate all statements directly as the experts would do.[20] This is an unreasonably strong requirement. It may appear intellectually conscientious to tell non-experts that they should not accept any statement unless they have evaluated its correctness for themselves, but that would deprive us of all benefits from the division of cognitive labor. The point of the division of cognitive labor is that we can accept statements based on the expert's counsel without conducting the relevant research for ourselves. It is sufficient for the acceptance of the expert counsel that we judge that those who provided the counsel are competent and honest.

Of course, the point remains that the basis of epistemic evaluation in meliorative personal epistemology is the epistemic resources available to each epistemic subject. Each epistemic subject must rely on their own epistemic resources to judge the (apparent) expert's competence and honesty. Even if there is some epistemic guideline for making this judgment, each epistemic subject needs to evaluate this guideline itself based on their own epidemic resources before applying it. The above point is only that the epistemic subject can judge the validity of statements indirectly from the competence and honesty of the experts who offer them, and it is through this indirect evaluation that the evaluation in meliorative social epistemology is connected to meliorative personal epistemology.

1.5 The Challenge of Cartesian Skepticism

It is pointed out in the previous section that meliorative epistemology comprises meliorative social epistemology and meliorative personal epistemology. This section sets the challenge of Cartesian skepticism in the two contexts. First, in the contest of meliorative personal epistemology, I argue against the conventional wisdom that the internalist constraints on epistemic evaluation inevitably lead to blanket skepticism. This is not much of a relief, it turns out, because it is meliorative social epistemology that provides us with informative general norms of epistemic evaluation. Moreover, well-known structural issues of traditional internalist epistemology—such as epistemic regress, epistemic circularity and empirically equivalent alternative hypotheses—reappear in meliorative social epistemology. Cartesian skepticism therefore remains a serious challenge in meliorative epistemology.

It appears obvious at first that we cannot overcome Cartesian skepticism under the internalist constraints on epistemic evaluation. How can an ordinary epistemic subject with limited epistemic resources solve the problem that has vexed highly capable epistemologists for centuries? Most people are unable or unwilling to follow—let alone construct—sophisticated arguments in support of the reliability of sense perception, memory or testimony. The expert's counsel does not help because epistemologists are in disagreement among themselves on many significant issues, and most people are unable to judge who are competent among the disagreeing epistemologists. It looks like the internalist constraints lead to blanket skepticism for almost all people. Blanket skepticism would be a serious problem for meliorative epistemology, not because it is counterintuitive, but because it would put all statements in one epistemic category, i.e. no statements are supported by the evidence and reasoning available to the epistemic subject. If there is no meaningful distinction between beliefs in good and poor epistemic standings, we cannot use epistemic evaluation to guide our epistemic practice to better results.

Fortunately, blanket skepticism is not an inevitable consequence of the internalist constraints. Those who fear blanket skepticism overlook a simple point about the internalist constraints, viz. the basis of epistemic evaluation is any evidence or reasoning that affects the epistemic evaluation of the statement—either positively or negatively.[21] It is true that the internalist constraints allow very limited evidence and reasoning to support the statement, but it also allows very limited evidence and reasoning to *discredit* the statement or *undermine* the support for the statement. For example, even if it is known to the relevant experts in the community that recently uncovered evidence disproves a widely accepted statement, it does not affect your evaluation of the statement under the internalist constraints unless the evidence becomes part of your epistemic resources. Similarly, even if there is a serious flaw in your reasoning and you would withhold your judgment on the conclusion if the flaw were explained to you, it is irrelevant to your epistemic evaluation under the internalist constraints unless the knowledge of the flaw becomes part of your epistemic resources. This makes the internalist constraints very *permissive*—your belief is in good epistemic standing even if there is decisive evidence against it or your reasoning is seriously flawed, unless the evidence or the knowledge of the flaw becomes part of your epistemic resources.[22] Once we see the permissive side of the internalist constraints, the specter of blanket skepticism dissipates.

There are, of course, some special cases where the internalist constraints are not so permissive, e.g. when the epistemic subject is an academic philosopher. She may be unable to escape blanket skepticism because she is aware of the forceful skeptical arguments against her beliefs, and is unable to respond to them with the epistemic resources available to her. But these are special cases. Most people are unaware of philosophical skepticism, and most people who are aware of it do not take it seriously because by their personal epistemic standard, philosophical skepticism is not a good reason for questioning their everyday beliefs. It may be suggested that the epistemic subject *should* be aware of philosophical skepticism and *should* take it seriously in order to improve her epistemic performance. But in meliorative personal epistemology it is the epistemic subject herself who makes that judgment. A third party's sense of epistemic norm is irrelevant. A small group of philosophers who worry about the consequence of the internalist constraints may think it is a general problem of meliorative personal epistemology, but they are merely imposing their own epistemic standards onto others against the internalist constraints of meliorative personal epistemology.

The internalist constraints are restrictive or permissive depending on different epistemic subjects with different epistemic resources. Some with rich epistemic resources may pay more attention to the danger of gullibility and thus maintain a high epistemic standard, but they may also have access to more evidence and reasoning to meet the high standard. Those with much less epistemic resources may have a lax epistemic standard and accept whatever appears true to them intuitively. There is no common standard of epistemic evaluation that is applicable across epistemic subjects.[23] The only general norm of meliorative personal internalism is that you (or any epistemic subject) should accept a statement just in case you have sufficient evidence for its truth *by your own standard*, though some standard is unacceptable to people who are better informed. Beyond this uninformative general norm, each epistemic subject is left alone and only encouraged to do her best with whatever evidence and reasoning available to her, but that is pretty much what we do anyway.[24] The general norm only reminds us that non-epistemic reasons—such as wishful thinking—should not be considered in the *epistemic* evaluation of the proposition.[25] Informative general norms are provided by meliorative social epistemology. It is for this reason that the primary focus in this book is meliorative social epistemology, and not meliorative personal epistemology.

In meliorative social epistemology the basis of epistemic evaluation is not limited to the epistemic resources available to each epistemic subject. It is the responsibility of the experts in the community to evaluate hypotheses in their fields of research for other members of the community. Those experts have access to the best empirical evidence and the best conceptual tools that are not available to ordinary epistemic subjects. Further, an expert in one field can turn to experts in other fields for help. In a community that takes full advantage of the division of cognitive labor, no single person—not even the best expert on the subject—thoroughly understands all of the evidence and reasoning that go into the evaluation of an expert statement. We cannot expect, for example, that a vision scientist understands how various devices in her lab, such as a motion tracker and an MRI scanner, work in complete detail. Among many other things, the vision scientist (qua vision scientist) will not be able to assess directly the robustness of the computer program that controls the devices. Just like anyone else in the community, an expert in one field turns to experts in other fields with which she is not thoroughly familiar, and she relies mostly on indirect evidence that those experts in other fields are competent and honest.

Interdependence among experts in the division of cognitive labor is a complex subject of social epistemology, and I can only make some basic remarks that are directly related to the project of this book. First, every element of epistemic evaluation in meliorative social epistemology must be recognized by some members of the community at some point.[26] So, there is a sense (an extended sense) in which epistemic evaluation in meliorative social epistemology is internalist, viz. the basis of epistemic evaluation is internal to the epistemic subject's community. I call this form of internalism *social internalism*. Note that "internal" to the community here means that the evidence or the reasoning must have been *actually* recognized by some members of the community at some point in time.[27] It is not enough that evidence or reasoning was "accessible" in the sense that it *would have been* recognized if one paid attention. Accessible evidence or accessible reasoning that has not been actually recognized by anyone in the community at any point in time plays no part in meliorative social epistemology. This is because the expert cannot endorse a statement that *would have been* supported by evidence and reasoning, which were never actually recognized by any member of the community. Mental states concealed from the expert's consciousness cannot be the basis of her epistemic evaluation, either.

Meliorative social epistemology is therefore strongly internalist: Every element of evaluation—every piece of evidence and every step of reasoning used either as the grounds for the evaluation or for showing the adequacy of the grounds—must be *actually* recognized by some members of the community at some points. Because of this "internalist" constraint on epistemic evaluation, meliorative social epistemology inherits the structural problems of traditional internalist epistemology. For example, experts in different fields of research cannot keep passing the buck of epistemic support indefinitely. Are there then *basic beliefs* held by some members of the community whose epistemic support does not depend on any other beliefs and thus provide the foundation of epistemic support? Perhaps *coherence* among the experts' beliefs makes them credible even if each belief has no credibility of its own? There has been some question of continuity between social epistemology and traditional epistemology because issues of social epistemology look more like subjects of social sciences.[28] That may be true in some form of social epistemology, but there is strong continuity between *meliorative* social epistemology and traditional epistemology: The former inherits the structural problems of epistemic support from the latter.[29]

Some people may find it odd to address Cartesian skepticism in the context of meliorative social epistemology. If we take Cartesian skepticism seriously, we cannot assume the existence of the natural world where the experts reside. The worry is that meliorative social epistemology may tacitly assume the truth of the natural world hypothesis. There is no need to worry. The BIV hypothesis is empirically equivalent to the natural world hypothesis, and if the BIV hypothesis is correct, then the virtual experts provide us with counsel that helps us navigate the virtual world. Just as the substance of scientific theories survive the demise of the natural world hypothesis as theories of the virtual world, the substance of the expert counsels survives the demise of the natural world hypothesis as counsels applicable to the virtual world. There is therefore no need in meliorative social epistemology to make the assumption that the natural world hypothesis is correct. However, to avoid unnecessary confusion, I am going to use a social version of the BIV hypothesis for the purpose of exposition.[30] According to this version of the BIV hypothesis, all members of the community, including the experts in various fields of research, are brains in vats kept alive and wired to a supercomputer that is programed to supply them with the kind of sensory input brains in skulls would receive from the natural environment.

In order to focus on the challenge of philosophical skepticism without distraction, I am going to simplify the setting of the discussion with the notion of the ideal evaluator. This evaluator has an unlimited access to the best empirical evidence and the best conceptual tools available to the community. In other words, the evaluator represents the totality of the community in terms of epistemic resources. The evaluator examines statements (both particular statements and epistemic guidelines) based on the best epistemic resources available to the community. If there is a disagreement on some subject matter among relevant experts in the community, the ideal evaluator would withhold judgment. If the relevant experts are epistemologists, and some of them raise legitimate and serious concerns about the structure of epistemic support for the reliability of our sense perception, then the ideal evaluator would withhold judgment on the reliability of our sense perception until the concerns are addressed properly. When the evaluator endorses a statement, it is indirectly the epistemic evaluation of beliefs held by members of the community. If the content of their belief is endorsed by the evaluator or the way the belief is formed is in accordance with the epistemic guideline endorsed by the evaluator, then the belief is in good epistemic standing in meliorative social epistemology. The evaluator's main task in this book is to determine whether she should endorse the natural world hypothesis to the community, or recommend that the community should revise its epistemic practice, e.g. to retreat to the modest goal of epistemic evaluation up to empirical equivalence.

Although the ideal evaluator has access to the best empirical evidence available to the community, there is no presupposition that the sense perception employed for collecting the evidence is reliable. So, the evaluator has access to the empirical evidence in the conditional form, e.g. that here is a hand if visual perception is reliable. Examining the reliability of sense perception is a big part of the evaluator's task. Meanwhile, the reliability of reasoning is not questioned in this book. Descartes (1984) introduced the evil demon hypothesis to cast doubt on our reasoning as well as our sense perception. However, I will assume in this book that formal results in logic and mathematics accepted by the experts in the respective fields are correct, and that the evaluator has access to these results. This means that the skeptic only questions the empirical ground for the statement. This is a big disclaimer, but Cartesian skepticism remains a formidable challenge. For example, in order to overcome Cartesian skepticism, the evaluator needs to provide empirical grounds for rejecting the BIV hypothesis in favor of the natural world hypothesis despite the fact that the two hypotheses are empirically equivalent to each other. The book takes up this challenge while taking full advantage of formal results in probability theory and statistical learning theory.[31]

This chapter described the subject matter of the book (Cartesian skepticism) and the method of inquiry adopted in the book (meliorative social epistemology). I want to reiterate a few points and the way they relate to the subsequent discussions. First, the distinction between meliorative epistemology and epistemography plays mostly negative roles in the subsequent discussions, viz. I will disregard objections that the views I propose have counterintuitive consequences. Whether or not the views are compatible with our pre-theoretical judgments is beside the point for the project of the book. Second, I will make use of any epistemic resources available to the community—both empirical and conceptual—but with no presumption against Cartesian skepticism. If the BIV hypothesis is correct, the resources help us understand the virtual world generated by the supercomputer. Third, meliorative social epistemology distinguishes the experts who evaluate beliefs and the epistemic subject whose beliefs are evaluated. This distinction plays a crucial role in Chapter 4 where a new Bayesian format of epistemic evaluation is introduced.

Notes

1 I am using the term "natural" in contrast with "supernatural" and "virtual" and not in contrast with "artificial". A chair is a natural object in this sense since it is an object and is neither supernatural nor virtual. This use is in line with the standard understanding of "naturalism" in contemporary philosophy.

2 Descartes himself uses the demon hypothesis to extend his skeptical challenge beyond empirical beliefs to all kinds of beliefs, including beliefs in logical and mathematical truths, but in this book I will restrict my attention to empirical beliefs about the natural world.

3 The term "meliorative" as a characterization of the broadly Cartesian approach in epistemology is due to Kitcher (1992, p. 64), who uses it in contrast with "analytic". Kaplan (1991) makes a similar distinction between meliorative epistemology (though he does not use the term) and "epistemology on holiday".

4 What this means will become clear in Chapter 6.

5 Interestingly, the idea of conceptual engineering is well recognized and influential in gender and race studies under the terms "analytical project" (Haslanger 2000), "ameliorative project" (Haslanger 2005; Jenkins 2016), and "engineering project" (Dembroff 2016). Its recognition and influence is understandable because merely describing the extant categories of gender and race may only reinforce current injustice many of the authors in the field wish to eliminate.

6 See Pust (2017) for a review of positions on the role of intuition in philosophical inquiries.

7 Gettier's second case is similar. The description of the situation makes it clear that Smith has a justified true belief that Jones owns a Ford or Brown is in Barcelona, but the description also makes it clear that Smith does not know this proposition.

8 Dreyfus (1972) stressed this aspect of human intuition in his polemics against the artificial intelligence community of the time. Dreyfus's main points have since been co-opted by the "sub-symbolic" approaches to artificial intelligence and cognitive modeling.

9 Pust (2001) uses this line of argument in his defense of philosophical intuition against its "explanationist" critics. According to Pust, the explanationist critics' argument against the use of intuition is self-defeating because one of its premises can only be supported by the use of intuition.

10 I am going to use Reichenbach's (1938) famous distinction in philosophy of science somewhat loosely as follows. The context of discovery is the context in

which hypotheses and objections to the hypotheses are generated, while the context of justification is the context in which we evaluate the generated hypotheses and objections to the hypotheses.

11 In the face of challenges to the use of intuition in philosophy, some philosophers have denied that analytic philosophers routinely rely on their intuition in their argument (Cappelen 2012; Deutsch 2015; Williamson 2008, among others). According to their interpretation, in prominent cases where intuition is thought to play a crucial role—such as Gettier's (1963) ten coins case, Foot's (1967) trolley case and Kripke's (1980) Gödel-Schmidt case—the authors do not actually use their intuition as evidence for or against the theories, but provide independent reasons for or against the theories. To put this point in the present context, the authors do not use their intuition in the context of justification, but only in the context of discovery. Not surprisingly, some philosophers challenge this interpretation (Climenhaga 2017), but I leave it to the historians of philosophy to determine the extent to which this interpretation is plausible. How this issue is resolved does not affect my point that meliorative epistemologists should not cite their intuition in the context of justification.

12 See Bowers et al. (1990) for an empirical study of the reliability of intuition in the context of discovery, i.e. how reliable intuition is in "generating hunches or hypotheses" (p. 73).

13 Cummins (1998) points out that unlike scientific instruments, philosophical intuition is not "calibrated" to test its accuracy.

14 Rawls himself states the importance of building a consensus as follows: "to serve as a public basis of justification for a constitutional regime, a conception of justice must be one that can be endorsed by widely different and even irreconcilable comprehensive doctrines. Otherwise the regime will not be enduring and secure" (Rawls 2001, p. 34).

15 There is no difference in this regard between "armchair" philosophy grounded in the intuition of a single philosopher and experimental philosophy (e.g. Knobe and Nichols 2008, 2013) grounded in the intuition of a group of people.

16 See Craig (1990) and Henderson and Greco (2015) for the proposal of analyzing the concept of knowledge from the perspective of its role in our life.

17 Smyth (2017) makes a similar point with regard to morality, viz. the evolutionary origin of morality should not be regarded as its present-day function because the relevant conditions that existed in our ancient history, such as the small size of the group, resource-scarcity and group instability (vulnerability to selfish individual actions), are mostly absent now.

18 Recall that the Cartesian skeptic does not question the substance of the currently accepted scientific theories, but only their interpretation. Scientists may not be studying physiological processes taking place in the natural world if the BIV hypothesis is correct, but their research survives as theories of the virtual world. So, the division of cognitive labor works fine even if the BIV hypothesis is correct—it allows us to construct better theories of the virtual world.

19 It is sometimes pointed out (e.g. Alston 1988) that we form most of our beliefs automatically in response to perceptual and testimonial input without conscious deliberation. This may appear to make epistemic guidelines largely irrelevant, but it is not necessary that we consciously think of the guidelines at the time of belief formation. We may accept some epistemic guidelines upon conscious deliberation and dispose ourselves to automatically form beliefs in the future in accordance with the guidelines.

20 Descartes states that his aim in writing *Discourse* is "not to teach the method which everyone must follow in order to direct his reason correctly, but only to reveal how I have tried to direct my own." (Descartes 1985b, p. 112) Presumably, each reader is expected to figure out the method of directing her reason for

herself. I will return to this theme shortly that each person is responsible for her own epistemology.

21 Alston (1986, p. 192) calls a factor that is relevant to the epistemic evaluation either positively or negatively "an epistemizer", while calling a factor that is positively relevant to the epistemic evaluation "a justifier". My point here, to put it in Alston's terms, is that the strong resource constraints of meliorative personal epistemology apply to epistemizers in general, and not only to justifiers.

22 Some people may find it counterintuitive that a *less* informed person is *more* likely to be in good epistemic standing. It is, of course, irrelevant to the project of meliorative epistemology whether the epistemic evaluation under the internalist constraints is in accord with our everyday intuition, but there is a simple explanation of the counterintuitiveness, viz. we usually evaluate people's belief from the third-person perspective, and not from the first-person perspective of the epistemic subject.

23 Goldman (1980, Section IV) describes a path of internalism that leads to epistemic relativism.

24 See Brandom (1998) and Stevenson (1999) for the view that beliefs and justified beliefs (nearly) converge from the first-person perspective. Foley (1993, p. 12) also notes that the "distinction between reporting reasons and endorsing them tends to collapse in the first-person".

25 The general norm may not play a significant role if the subject's limited epistemic resources do not allow her to distinguish wishful thinking and evidence-based beliefs clearly.

26 This is only a necessary condition. It is required further that the relevant evidence and reasoning must be communicated properly among the members of the community, and that all the elements must be put together in a structurally sound way.

27 This is the meaning of "internal" in the internalist restriction for the purpose of social internalism.

28 Alston notes in regard to Goldman's influential work in social epistemology (Goldman 1999) that "much of the material in Goldman's book would be rejected by many contemporary epistemologists as "not real epistemology", and relegated to sociology, social psychology, or other social sciences, or perhaps to the philosophical foundations thereof" (Alston 2005, p. 5).

29 Some meliorative social epistemologists, who are externalists in the traditional sense, take up the structural problems of epistemic support. For example, Goldman (1999, Section 3.3) defends Bayesian coherentism as a (qualified) solution to the problem of circularity.

30 See Veber (2015) for a social version of the BIV hypothesis.

31 This is often what actually takes place in traditional internalist epistemology, viz. the evaluator takes up the challenge of philosophical skepticism while taking full advantage of sophisticated conceptual tools that are not available to ordinary epistemic subjects.

2 Truth as Correspondence

2.1 Correspondence Theory vs. Deflationism

Meliorative epistemology aims to guide our epistemic practice to better results, and better results means increasing true beliefs and avoiding false beliefs when our judgment is qualitative, and making the probability as close to the true value as possible when our judgment is quantitative. However, there are different conceptions of truth, and how we respond to Cartesian skepticism depends in part on the way we conceive of truth. It is therefore necessary to examine the different conceptions of truth, especially the two major conceptions in contemporary philosophy: the correspondence theory of truth and the deflationary theory of truth.

The traditional conception of truth is correspondence, i.e. truth consists in correspondence with reality. The correspondence theory of truth comes in various forms but we can put the basic structure as follows. Beliefs and statements represent reality in various ways. True beliefs and true statements correspond to reality in the sense that they represent reality the way it actually is, while false beliefs and false statements represent reality the way it actually is not.[1] In many versions of the correspondence theory beliefs and statements have *components* that relate to different *constituents* of reality. For example, a singular term represents (refers to) an object, and a predicate represents (refers to) a property.[2] An atomic statement that consists of a singular term and a predicate is then true if and only if the object represented by the singular term exemplifies the property represented by the predicate. Similarly, a belief consisting of a singular mental token and a general mental token is true if and only if the object represented by the singular mental token exemplifies the property represented by the general mental token.[3]

The correspondence theory of truth makes the challenge of Cartesian skepticism ostensibly intelligible. Most of our beliefs about the natural world should be false if we are brains in vats.[4] However, not all forms of the correspondence theory make Cartesian skepticism a threat. Suppose, for example, someone adopts the policy of charitable interpretation and assigns representational contents to beliefs (statements) in such a way that as many beliefs (statements) as possible come out true.[5] The policy appears sensible but unless it is suitably qualified, the resulting interpretation makes most beliefs held by a brain in a vat come out true because, by the most charitable interpretation, the BIV's beliefs represent various aspects of the virtual world generated by the supercomputer, instead of aspects of the natural world. The natural world interpretation uncharitably makes most of the BIV's beliefs false.

The Cartesian skeptic, or anyone who takes Cartesian skepticism seriously, would reject this easy way out. But then what is the appropriate representational relation for interpreting the BIV's beliefs? It is hard to spell it out, and it is even harder to explain the epistemic reason for adopting the spelled-out relation for interpreting beliefs and statements. Similar difficulties arise for any substantive theory of truth. For example, the advocates of the coherence theory of truth must spell out the relation of coherence, and explain the epistemic reason for adopting the spelled-out relation for evaluating beliefs and statements by that relation. As the enormity of the challenge becomes evident for any substantive theory of truth, the deflationary theory of truth looks attractive in comparison. According to the deflationary theory, the role of truth is exhausted by the simple schema of equivalence: The belief or the statement that *p* is true if and only if *p*. There is no need to spell out the relation of representation, coherence, etc. The concept of truth plays no substantive role.

Most deflationists acknowledge that truth has a technical function of expanding the expressive power of the language and the mental states. For example, the predicate "is true" allows me to say that everything I stated at the meeting is true, without specifying or even remembering exactly what I stated at the meeting—e.g. I may be so confident of my epistemic competence and honesty that I cannot imagine stating anything false. The concept of truth is therefore not entirely redundant, but I would be able to say the same by exhaustive enumeration—i.e. if I stated at the meeting that snow is white, then snow is white; if I stated at the meeting that grass is green, then grass is green. Of course, no one can enumerate infinitely many conditional claims, but the point is that truth is only a formal device for expanding the expressive power of the language. The same point applies to the concept of truth used in thoughts. It allows me to think that everything I believed on the subject matter is true, without specifying or even remembering exactly what I believed on the subject matter, but again truth is only a formal device for expanding the expressive power of the mental states.

The equivalence schema is not controversial. Though some substantive theories of truth are difficult to reconcile with the equivalence schema,[6] the correspondence theory of truth has no trouble explaining why it holds, viz. if the belief has the representational content that *p*, then the belief is true if and only if *p* (i.e. if and only if the belief represents the world the way it actually is). For example, if the mental tokens <snow> and <white> are in the substantive relation of reference to snow and whiteness, respectively, then the belief that snow is white is true if and only if snow is white (if and only if the belief that snow is white represents the world the way it actually is). The equivalence schema is therefore uncontroversial from the perspective of the correspondence theory.

The equivalence schema itself is then neutral between the two theories of truth, but the deflationist uses it to avoid the challenge of spelling out the relation of representation, which is supposed to determine the way reality must be for a belief or a statement to be true. The deflationist asks: If we can eliminate the truth talk by the equivalence schema, why do we need to spell out the relation of representation? We can distinguish two versions of deflationism here. One version takes deflationism to be simply a way of eliminating the truth talk.[7] It is not necessary for this version of deflationism to repudiate the correspondence theory of truth. Even if the correspondence theory of truth is correct and the equivalence schema holds because *p* is the representational content of the belief that *p*, there is no need to spell out the relation of representation if we can drop the truth talk by the equivalence schema. I call it "neutral deflationism" because it leaves the question open whether the content that *p* should be understood representationally.[8]

The second version of deflationism, which is more radical, rejects the representational understanding of the content that *p* and thus the correspondence theory of truth. We can see how it differs from neutral deflationism in the case of a foreign sentence. We may ask, for example, under what condition the French sentence "Voici une main" is true. If you are a monolingual English speaker, you cannot say it is true if and only if voici une main. You can mention the French sentence and ascribe truth to it—correctly or incorrectly—but the right-hand side of the equivalence schema makes no sense to you. There is a sensible answer to the question, viz. the French sentence "Voice une main" is true if and only if here is a hand because "Voice une main" and "Here is a hand" have the same representational content, and "Here is a hand" is true if and only if here is a hand. However, this option is not open to the radical deflationist who rejects the representational understanding of the content that here is a hand.

The radical deflationist can still accept that "Voici une main" and "Here is a hand" have the same content, but not the same *representational* content. Similarly, two people's cognitive states may have the same content, but not the same *representational* content.[9] This means that a belief—or a cognitive state if we reserve the term "belief" for those with representational contents—is individuated by its non-representational role, e.g. its computational role in the person's psychology. The computational role of a cognitive state is defined by its causal relations to other mental states, sensory inputs, and behavioral outputs—with no reference to its representational content. According to this understanding, the content of the cognitive state, that *p*, is the computational role it plays in the person's psychology. The same point also applies to a statement. The radical deflationist denies that a statement has a representational content. Instead the content of a statement, that *p*, is its non-representational role, e.g. its conceptual role. The meaning of a statement then consists in the way it relates to other statements, sensory inputs, and behavioral outputs.[10]

Some people may feel that the view just described is a theory of contents and meaning, and not really a theory of truth. It may be suggested that only "neutral deflationism" is the deflationary theory of truth, though the deflationist is free to combine it with the non-representational analysis of contents and meaning.[11] I do not object to this use of the term "deflationism", but the point of substance, for the present purpose, is that the deflationist can avoid the challenge of spelling out the relation of representation only by adopting a non-representational analysis of contents and meaning. If the belief has the representational content that *p*, then even if we drop the truth talk by the equivalence schema, the question whether *p* or not *p* amounts to the question whether reality is the way the belief represents it or not, and we need to spell out the relation of representation to answer the question.[12] So, deflationism is an interesting theory, for the present purpose, only in combination with the non-representational account of contents and meaning. I will therefore focus on this view regardless of its label, i.e. whether it is called radical deflationism or deflationism with the non-representational account of contents and meaning.

The problem of truth—for the purpose of this book—is now clear, but the problem raises a difficult methodological question. Meliorative epistemology aims to guide our epistemic practice to better results, where the better results are understood alethically. So, competing proposals in meliorative epistemology are evaluated in relation to the alethic goals, but since the problem of truth concerns the nature of the alethic goals themselves, we cannot adjudicate the two conceptions of truth in relation to the alethic goals whose nature is in dispute. Recall also that pre-theoretical judgments play no role in meliorative epistemology. For example, it may (or may not be) part of our pre-theoretical understanding that truth is a bivalent property so that either the statement that *p* or the statement that not *p* must be true, but we may drop this condition in meliorative epistemology if it serves no epistemic purpose. If we set aside pre-theoretical judgments, how can we adjudicate the two conceptions of truth? I will try two approaches. First, I examine whether it is a consequence of the resource constraints in meliorative social epistemology that we assign representational contents to statements and beliefs (Section 2). I will then examine whether it serves the alethic goals to assign representational contents to statements and beliefs, where truth in the alethic goals is truth of neutral deflationism with no commitment to either of the two conceptions of truth (Sections 3 and 4).

2.2 Reference by Deference

This section examines an influential view of reference, according to which some types of linguistic expressions represent (refer to) constituents of reality via the social relations of deference. If this is correct, then statements of which these expressions are components have the representational contents that go beyond their conceptual roles in the speaker's psychology. The proponents of the view motivate and defend the relations of deference mostly by pre-theoretical judgments in thought experiments. Though meliorative epistemology is not in the business of accommodating pre-theoretical judgments, I want to investigate whether the proposed relations are necessitated by the resource constraints in meliorative social epistemology, and whether they can help us resolve the dispute over truth.

I want to begin with Hilary Putnam's proposal of the division of linguistic labor. Putnam (1975) describes a person who is aware that the terms "beech" and "elm" refer to trees of different kinds, but who cannot distinguish them for himself. Putnam argues that the references of these terms uttered by this person are determined by the way the experts in the community use them. This is why the terms "beech" and "elm" uttered by this person refer to beeches and elms, respectively, despite his inability to distinguish them. It is a form of content externalism because the references of the terms depend on the conditions external to the speaker. In a community where the botanists use the terms "beech" and "elm" in reverse to refer to elms and beeches, respectively, the references of the terms uttered by the ill-informed person are also reversed even if his intrinsic properties remain exactly the same.

Tyler Burge (1979) makes a similar point with regard to mental contents. Burge's case involves a person who is unaware that arthritis is a disease of joints and believes falsely that he has arthritis in his thigh. The point of the case is that the reference of this person's mental token <arthritis> is determined by the way more knowledgeable members of his community use the term "arthritis". This is why the person's belief that he has arthritis in his thigh is false regardless of the conditions in his thigh. However, if this person belongs to a community where the medical experts use the term "arthritis" to refer to rheumatoid disease in general, his belief may be true even if the person's intrinsic properties remain the same.

The basis of Putnam's and Burge's arguments is pre-theoretical judgments in thought experiments and their judgments seem to be shared by many philosophers. Although the pre-theoretical judgments play no role in the evaluation of theories in meliorative epistemology, this type of content externalism due to deference to the experts makes good sense in the context of meliorative social epistemology. Recall that epistemic evaluation in meliorative social epistemology is not constrained by the limited resources available to the individual epistemic subject. It is only constrained by the much richer resources available to the epistemic subject's community. It is the appropriate experts in the community who conduct the evaluation and disseminate the results to the members of the community. The division of cognitive labor implies that those members of the community who are less informed should defer to the appropriate experts both on the substance of the evaluation and the way the evaluation is conceptually articulated. When some non-experts in the community make a statement, or form a belief, that involves concepts developed by the experts, the statement and the belief should be interpreted by the way the experts employ the concepts. Content externalism due to deference to the experts is therefore an important part of meliorative social epistemology.

It is possible to extend the idea of reference by deference further. One such extension is deference to the future experts. Sometimes even the best experts in the community only have a superficial grasp of the subject matter. This was common in the past, especially about natural kinds such as water and gold, and there are still some cases of deference to the future experts. For example, even the best astrophysicists are unsure about the nature of *dark matter* at this point. Given this state of research, it makes good sense for the experts not to define the term "dark matter" based on the limited information that is currently available, but to defer to the future generation of scientists who are likely to uncover its identity based on better epistemic resources. So, the term "dark matter" as it is currently used refers to whatever substance the future scientists will uncover and refer to in explaining (most of) the superficial (by the standard of physical sciences) features currently associated with the term. The mental token <dark matter> as it is currently used also refers to the substance in the same way by deference to the future scientists.

Some people may worry that deference to the future experts is at odds with the resource constraints of meliorative social epistemology, according to which epistemic evaluation must be based on the epistemic resources available to the community to which the epistemic subject belongs. The worry is that we cannot improve our epistemic practice now by the future experts' evaluation that has yet to come. However, deference to the future experts does not mean that we turn to the future experts now for their help, which is obviously impossible. It means that we withhold judgments till more epistemic resources become available. This is appropriate not only on the substance of evaluation, but also on the way the evaluation is conceptually articulated. Where there is insufficient information available, we may only loosely describe the concept for now with the anticipation that the future experts will be able to define it properly based on their better epistemic resources.

In most cases of natural kinds, the current experts already have sufficient information for their proper definition. So, there is no good reason for deference to the future experts. Of course, there may be some unexpected discovery that prompts their revision in the future, but that is no different from revision in our epistemic evaluation. Where there is strong evidence, we need not withhold epistemic judgment even if there is a small chance of future retraction due to some unexpected discovery. At a certain point we consider the case to be sufficiently strong for acceptance and dissemination. Similarly, we define a concept at a certain point based on available information even if we may be compelled to reconsider it in the future.

It is also possible to extend the idea of reference by deference to proper names, such as "Gödel" and "Feynman". According to Kripke's (1980) influential account, the reference of a proper name is determined by a chain of communication. For example, if I learned the name "Gödel" from my classmate, my utterance of the name "Gödel" refers to whatever my classmate's utterance of the name "Gödel" refers to. If the classmate in turn learned this name from her professor, her utterance of the name "Gödel" refers to whatever that professor's utterance of the name "Gödel" refers to, and so on. This form of deference also makes good sense from the standpoint of meliorative social epistemology when we do not have much information for identifying the referent, as is often the case with proper names. Of course, there are usually no experts in the community who specialize in a particular object that a proper name refers to. So, when we have little information on the referent of a proper name, we simply defer to the person from whom we learned the proper name. This person may also have little information on the referent, in which case she may defer to the person from whom she learned the proper name, and so on. There is no need in my view to go all the way back to baptism where the proper name was introduced. The chain of deference stops with someone with sufficient familiarity with the referent. The chain may not even start if we are sufficiently familiar with the referent. The same points apply to singular mental tokens such as <Gödel>, <Feynman>, etc.

So, the resource constraints in meliorative social epistemology support reference by deference about technical terms, natural kind terms, and proper names. In general each member of the community is expected to use the language the way other members of the community use it. Otherwise, we cannot rely on testimonial evidence. Where there is some difference among members, the basic norm is to follow the majority, but in cases where some members are much better informed on the subject matter than the majority, it is sensible to defer to these members. Deference to the experts extends to the interpretation of mental states as in Burge's arthritis case where the mental token <arthritis> is firmly tied to the technical term "arthritis" in the public language. However, we need some caution not to overstate the case. Reference by deference is a sensible social norm, but when it comes to explaining and predicting the subject's verbal and non-verbal behavior, only the way the subject understands the term is relevant. For example, when a patient uses medical terms such as "cancer" and "diabetes", the doctor should not interpret them by the professional standard. When it comes to explaining and predicting the patient's verbal and non-verbal behavior, the doctor should assume that the patient's understanding is constrained by her limited personal epistemic resources on the subject.

With this proviso, it is sensible to incorporate reference by deference into the framework of meliorative social epistemology. The question now is whether reference by deference helps us settle the dispute over truth. It may seem that the answer is yes. The relations of deference look like substantive relations between components of a statement and constituents of reality, which the deflationist tries to avoid in favor of non-representational relations. For example, the terms "beech" and "elm" refer to beech and elm, respectively, by deference to the experts, even if their non-representational roles in the speaker's psychology are indistinguishable. However, a little reflection reveals that reference by deference does not support the representational account of contents and meaning.

First, in addition to the equivalence schema on truth, the deflationist can also turn to the simple schema of reference: The term ⌜*t*⌝ refers to *t*. The deflationist can then maintain that the role of reference is exhausted by this schema. There is no need to spell out the relation of reference because we can drop the notion of reference by uttering ⌜*t*⌝, instead of ⌜what ⌜*t*⌝ refers to⌝ and analyze the meaning of the term by its conceptual role. Note further that the idea of reference by deference is neutral on the nature of reference. It only requires that we defer to the experts for the proper use of certain terms, where the proper use may be determined by their representational relation to some constituents of reality, or by their non-representational role in the experts' conceptual system. If reference turns out to a representational relation, at least for some terms, then the statements involving these terms must also have representational contents. However, there is no way of telling just from reference by deference whether statements made by the experts have representational contents.

In short, reference by deference only moves the location of the dispute from the contents of statements in general to the contents of the statements made by the experts. We must resolve the dispute over truth in some other way. To avoid unnecessary complication and distraction, I will assume in the following discussion that the epistemic subjects follow the social norms in their use of the language. Those who dislike this assumption can understand the discussion as applicable to the experts to whom we defer.

2.3 Mental Representation

So, there are no answers yet to the questions whether it is helpful in some way to assign representational contents to beliefs and statements, and what the representational relation should be like to deliver the maximum benefit if the answer is yes. One place to seek answers to these questions is the philosophy of psychology, where the role of representational contents has been a major topic. The discussion unfolds, very roughly, in the following way. The proponents of intentional (representational) psychology argue that assigning representational contents to people's mental states helps us explain and predict their behavior. It is a common practice to explain people's behavior by their beliefs and desires whose representational contents are indicated by that-clauses, e.g. we may explain why Mary stays home by her belief *that it is snowing* and her desire *that she stay home when it is snowing*.[13] The ascription of beliefs and desires supported by successful explanations of behavior in turn helps us predict people's behavior. Successful prediction means that the prediction is *true*, but it is not necessary to decide the nature of truth with regard to the prediction. We can adopt neutral deflationism about the truth of the prediction while focusing on the issue of whether to assign representational contents to the mental states of the person whose behavior is predicted. If the answer is yes, the person's mental states (beliefs) can represent reality correctly, and thus they can be true in the sense of correspondence with reality.

The opponents of intentional psychology reject the explanation of the subject's behavior by the representational contents of her beliefs and desires. They prefer a psychological theory with no representational contents assigned to mental states. Computational psychology is an obvious candidate, where the explanation and prediction of people's behavior is based on the non-representational roles of their mental states as characterized by their relations to other mental states, sensory inputs and behavioral outputs. There is no need in computational psychology to assign representational contents to mental states.

Computational psychology has some strong initial points in its favor. For one thing, the computational approach in cognitive science produced significant discoveries, while intentional "psychology" used in everyday life seems to be mostly platitudes with no surprising discoveries. Some argue for this reason that intentional psychology is "folk psychology" to be replaced by cognitive science that takes the computational approach.[14] A further point against intentional psychology is that representational contents are often too coarse even for the purpose of folk psychology. For example, two mental tokens <the morning star> and <the evening star> denote the same object, Venus, so there is presumably no difference in the representational contents between the belief that the Morning Star is the Morning Star and the belief that the Morning Star is the Evening Star, but their roles in people's psychology are often very different and they may lead to different predictions of behavior.[15] This means that even the advocates of intentional psychology may need some non-representational elements, such as computational roles, in their comprehensive psychological theory. If so, the real question is not whether we need non-representational psychology—it seems we do—but whether we *also* need intentional psychology.

There is also the enormous challenge of spelling out the representational relation in intentional psychology. Unless the subject puts her beliefs and desires in words—and we understand her language, and we trust her competence and honesty—we must identify her beliefs and desires by her sensory inputs and behavioral outputs since we cannot directly observe her mental states. Further, even if we manage to assign representational contents to them, and their representational contents somehow help us explain and predict the subject's behavior beyond computational psychology, it is still unclear what is the advantage for the subject herself when her beliefs are true, as determined by the representational contents assigned for explanatory and predictive purposes. In other words, it is unclear why the subject should aim at increasing true beliefs and avoiding false beliefs.

In short it is difficult to explain why we should assign representational contents to beliefs and desires. However, there is one area where intentional psychology is compelling, viz. psychology of sense perception. Unlike beliefs and desires, it is easy to tell what the subject is seeing, hearing, etc. in a given situation, independently of her other mental states and in the absence of verbal expression. We can even tell what a non-human animal is seeing, hearing, etc. in a given situation. The representational content of a sensory state is essentially its normal distal cause. For example, where there is a hand in front of the subject with open eyes, we ascribe to her a visual state that represents a hand in front of her. Of course, it would be nice if we could identify the *proximal* cause (sensory stimulus) of the visual state for fine-grained analyses, but that is not realistic outside the laboratory setting. Also, there is an indirect line of reasoning for the claim that an accurate representation is advantageous for the subject herself.[16] If one assumes, as is reasonable, that the psychological system has evolved to produce appropriate behavior under normal conditions, then the subject's behavior is likely to be appropriate when the actual cause of her sensory state is its normal distal cause. So, we can make a strong case for intentional psychology in the area of sense perception.

Moreover, once intentional psychology finds a foothold there, we can extend it further to assign representational contents to beliefs through their non-representational relations to sensory states. To put it schematically, let S_1, ..., S_n be sensory states with the representational contents $R(S_1)$, ..., $R(S_n)$, respectively, and B_1, ..., B_m be beliefs. Suppose the non-representational (e.g. computational) relation NR holds among S_1, ..., S_n, B_1, ..., B_m in the person's psychology while the non-representational (e.g. nomological) relation NR* isomorphic to NR holds among $R(S_1)$, ..., $R(S_n)$, $B_1{}^*$, ..., $B_m{}^*$ in the world. We can then assign $B_1{}^*$, ..., $B_m{}^*$ to the beliefs B_1, ..., B_m, respectively, as their representational contents, $R(B_1)$, ..., $R(B_m)$.

This is all promising, but two major problems remain about the normal distal cause. First, its relation to the sensory state is causal in nature, just like any relations in computational psychology among mental states, sensory inputs and behavioral outputs. So, we can include normal distal causes in the causal network as an extension of computational psychology. This allows us to provide thoroughly causal explanations and predictions of behavior without representational relations. It is unclear why we must accord a special status to the causal relation in perception and mark it as *representational*. Second, it is also unclear what it means for the distal cause of a sensory state to be normal. If we are BIVs whose sensory states are generated by a supercomputer, what are their normal distal causes? One possible answer is some configurations in the supercomputer. But then the BIV's sensory states are mostly accurate. This seems odd, but what else can be their normal distal causes? It is not enough to formulate some notion of the normal distal cause that makes the BIV's sensory states mostly inaccurate. We need some reason for adopting that notion beyond pre-theoretical judgments that are not relevant to the project of meliorative epistemology.

To answer these questions, I propose to approach the representational relation in sense perception from the perspective of the epistemic subject herself, instead of the third-person perspective as a relation between someone else's sensory state and its normal distal cause. The first-person approach may appear puzzling or even doomed because the representational relation is a relation between two relata, e.g. the epistemic subject's sensory state and some conditions in the environment, while the epistemic subject's sensory experiences only reveal one end of the relation.[17] It is therefore doubtful that we can gain any useful insight on the relation of representation from the epistemic subject's first-person perspective. There is, however, an attractive response to this doubt, viz. it seems the subject can tell from her first-person perspective the conditions of the environment that her sensory experience represents. I am going to examine whether this response is consistent with the basic premise of representationalism that sensory representation is a relation between two relata—the sensory experience and some condition in reality.

There is one idea that I want to set aside first. It is tempting to argue that accurate sensory representation is an *internal relation* in the sense that whether the relation holds or not is solely dependent on the intrinsic properties of each relatum. To illustrate an internal relation, if you are six feet tall (your intrinsic property) and your brother is five feet tall (his intrinsic property), then *ipso fact* you are taller than your brother. Since no information beyond their intrinsic properties is needed to establish their relation, *being taller than* is an internal relation.[18] Similarly, it might be suggested that whether or not our sensory experience accurately represents reality is solely dependent on what the sensory experience is like in itself and what reality is like in itself. It may seem, more specifically, that the sensory experience accurately represents reality if and only if the two relata are isomorphic to each other. It seems, then, that accurate representation is an internal relation since isomorphism is an internal relation.

The suggestion may seem plausible initially, but there is a major problem. To make the problem easier to see, consider sensory experience whose representational content is existentially quantified, e.g. the visual experience that here exists a hand.[19] The representational content is then a property exemplified at a certain location in space.[20] The problem is that this location in space is specified in relation to the sensory state. In order for my visual experience that here is a hand to be accurate, there must be a hand in front of the body that houses my sensory state. It does not count if there is a hand in some faraway place. In other words, accurate sensory representation requires that a proper spatial relation holds between the sensory state and the location of exemplification, and the spatial relation is not an internal relation, e.g. whether or not a hand is in front of my body is not solely determined by the intrinsic properties of my sensory state and the hand. This means that accurate sensory representation is not an internal relation.

Let us return to the idea that the subject can tell from her first-person perspective the conditions of the environment that her sensory experience represents. The idea looks suspect once we reject the suggestion of an internal relation. In order for the subject to tell from her first-person perspective the conditions of the environment that her sensory experience represents, the representational contents must be supervenient on the subject's sensory states, but if the relation of sensory representation is a substantive (spatial) relation, and not an internal relation, then what conditions her sensory experience represents depends on the way her sensory state relates to her environment. How can the representational contents, then, be supervenient on the intrinsic properties of the subject's sensory states?

The answer to the question is that the relation of representation is itself revealed by our sensory experience, i.e. we can tell from our own sensory experience where the perceived object must be located relative to our body in order for the sensory experience to be an accurate representation of the environment. If the relation of representation is itself revealed by our sensory experience, then the representational contents of sensory states can be supervenient on the intrinsic properties of the sensory states even if the relation of representation is a genuine (non-internal) relation between the two relata. Note also that our sensory experience of seeing a hand misrepresents reality in the skeptical scenario, according to this view, because there is no hand in the proper spatial location relative to the BIV that houses the sensory state.

2.4 Representational Parsing of the Visual Input

The discussion in the preceding section is promising: It seems we can grasp from the first-person perspective the representational relation that holds between our own sensory experience and the condition of the environment. However, the real task in meliorative epistemology is to investigate whether there is good epistemic reason for understanding the sensory experience in this way, and I address this question in this section.

In light of the importance of the spatial relation in the specification of the representational content of the sensory experience, my focus in this section will be visual experience because visual experience represents spatial properties in the clearest way among all sensory modalities.[21] I am going to defend the following two claims about our visual experience. First, the spatial relation between the visual state and the perceived object—a relation that is itself represented by the visual experience—plays a crucial role in constructing the three-dimensional model of the natural world from the two-dimensional visual input. Second, having the three-dimensional model of the natural world is helpful in successfully predicting the visual experience to come. The upshot is that the representational parsing of the visual input improves our epistemic performance. Successful prediction means that the prediction is *true*, but as mentioned already, it is not necessary to decide the nature of this truth in advance. We can adopt neutral deflationism about the truth of the prediction while focusing on the issue of whether assigning representational contents to our own visual states helps us make true (in the conception-neutral sense) predictions.

Brief remarks are in order on what is *not* claimed here. First, it is not part of my claim that the epistemic subject is aware of the way the representational relation helps the construction of the three-dimensional model of the natural world, or the way the resulting three-dimensional model improves our epistemic performance. The process is mostly subpersonal, and the analysis below describes the benefit of the mostly subpersonal process. Second, I am not making any specific claim about the process that takes place at the level of an algorithm, let alone at the level of neural implementation, through which the two-dimensional visual input is transformed into the three-dimensional model of the natural world. I only argue that the transformation, whatever its actual process turns out to be, hinges on the spatial relation between the visual state and the perceived object, which is revealed by our visual experience.

Here is a brief overview of the structure of our visual space. Our everyday conception of space is three-dimensional, i.e. we take it for granted that we are seeing some aspects of the three-dimensional world. However, it is also clear on some reflection that our phenomenal visual space (or simply "visual space") is not three-dimensional. It is customary to say that we see a hand, but we actually see only the facing side of a hand. We do not see its backside, let alone its inside, which is certainly part of the three-dimensional world. It is our visually informed conception of the world, and not our visual space itself, that is three-dimensional. This observation tempts us to think that our visual space is two-dimensional, but that is not accurate either because we actually see a configuration of surfaces with varying depths. These surfaces do not occupy a two-dimensional space, not even a warped two-dimensional space, because they often have discontinuous depths—one surface may be close to us while an "adjacent" surface may be far away. Our visual space is therefore "two-and-a-half dimensional" to use David Marr's (1982) term. So, neither our visually informed model of the world nor our visual space is two-dimensional. However, the visual input we receive is two-dimensional. Photoreceptors in the retina do not distinguish depths. They only register the wavelengths and intensities of the lights that arrive on the two-dimensional surface of the retina. Our visual perception therefore involves three different spaces. We receive 2-D visual input, and we convert it into a 3-D model of the world, in which our 2.5-D visual space is embedded.[22] The first claim of this section is that the representational relation plays an important role when we parse the 2-D visual input in this way.

There are a variety of ways to construct a 3-D model from 2-D visual input. Some argue that we connect vision to other sensory modalities in the construction, e.g. Berkeley (1948) famously argued that the 2-D visual input indicates depth by its customary connection with touch. However, there are many depth cues in vision itself such as relative sizes and occlusion. Some of them are not suitable for the present purpose since they need to be combined with some prior knowledge of the natural world, but others are reasons on their own for parsing the 2-D visual input into a 3-D model. Of these I will focus on vision in motion where it is easy to see the role of the representational relation in the construction of the 3-D model.[23]

The first thing to note about vision in motion is that 2-D visual input we receive changes quickly and drastically. A sudden global lateral shift occurs when we turn around. As we move forward, a global centrifugal shift takes place, i.e. the entire configuration of the visual field expands outward, forcing its peripheries to disappear. We are usually oblivious to these changes in the 2-D input because they are automatically treated as a result of change in the position of the visual sensor (the eye). A sudden global lateral shift is automatically attributed to a redirection of the visual sensor; a global centrifugal shift is automatically understood as a result of the forward movement of the visual sensor.[24] In short we interpret our 2-D visual input in motion as a product of the dynamic relation between the visual apparatus (of which the visual sensor is the front part) and the environment. The tacit understanding here is that the visual apparatus collects information from the environment through the visual sensor as the latter changes its position (location and direction), so that the 2-D visual input we receive reflects both the way the environment is and the position of the visual sensor in the environment.[25] The result of this parsing of the 2-D visual input is a 3-D world with a directed center, i.e. we obtain a 3-D model of the world with the visual sensor at its center in a certain direction. The 3-D world as such is considered perspective free and mostly stable, but as the visual sensor changes its location and direction, we obtain rapidly changing 2-D visual input.

So, the flux of the 2-D visual input is parsed into a relatively stable 3-D world with a directed center. The 2.5-D visual space, which consists of surfaces with varying depths, is then embedded in the 3-D world. The surfaces with varying depths are taken to be the facing sides of the nearest non-transparent objects relative to the directed center of the 3-D world.[26] The resulting visual experience is a 2.5-D representation of the 3-D world from its directed center, and its representational content consists of the facing sides of the nearest non-transparent objects with varying depths. In this parsing of the visual input, the visual state plays two roles. From the first-person perspective, it is a visual experience that represents a 3-D world. However, the visual state also belongs to the represented 3-D world because in order to receive the visual input, the visual state must be connected to the visual sensor that belongs to the represented 3-D world.[27] The visual state is therefore part of the world it represents.[28] To play the two roles at the same time, the visual state cannot be simply a nexus in the causal network; it must be representational.

The second claim of the section is that the 3-D parsing of the 2-D visual input is helpful in predicting the visual experience to come. The 3-D parsing allows us to maintain the view of the mostly stable environment in the face of frequent and drastic changes in the 2-D visual input, and this in turn makes it easier to predict the visual experience to come. For example, we expect houses and trees to remain where they are even when we are away or are looking elsewhere, and these expectations are fulfilled most of the time when our visual sensor returns to its original position.

Of course, not all changes in the visual input are caused by changes in the position of the visual sensor. Some changes reflect changes in the environment, but it is not difficult to distinguish them from those caused by changes in the position of the visual sensor. Local changes in an otherwise stable visual space indicate changes in the environment, while global systematic changes indicate changes in the position of the visual sensor. In some cases both kinds of changes take place at the same time—e.g. when we chase a running dog in the yard. In such cases changes in the visual input reflect both changes in the environment and changes in the position of the visual sensor at the same time. But even in such cases the stable background—such as the ground, the trees and the fences—allows us to differentiate the two kinds of changes fairly easily. It is against the stable background that we recognize the changing part of the environment—e.g. the dog's location in the yard. This distinction is helpful, obviously, in predicting the visual experience to come.

The case of short-term occlusion also illustrates how we benefit from the representational parsing of the 2-D visual input. Suppose a moving surface of a bus temporarily eclipses the surface of a mailbox across the street. We acknowledge this with no sense of surprise and remain confident that the mailbox is still where it was. The surface of the mailbox has disappeared from the 2.5-D representation simply because the mailbox is no longer the nearest non-transparent object in the direction of the visual sensor. When the bus moves on, the surface of the mailbox reappears since it is again the facing side of the nearest non-transparent object in the direction of the visual sensor. We have a view of the world that is mostly stable with many unperceived objects, and it is the representational parsing of the 2-D visual input that allows such a view. In the absence of representational parsing, any object that goes out of sight simply ceases to exist, and that makes it hard to make accurate predictions of the visual experience to come. The representational parsing of the 2-D visual input therefore serves the epistemic goal of increasing predictive success and avoiding predictive failure.[29]

These points about visual perception—the 3-D parsing of the 2-D visual input and the advantage of the stable 3-D model of the world—are not controversial in vision science. I went over them to draw attention to their implication in the dispute over truth with an emphasis on the role of the representational relation from the first-person perspective. These points seem underappreciated in the literature on the representational content, where it is common to take a third-person approach and ask what representational content, if any, we should assign to someone else's mental state.

As mentioned earlier, some puzzles remain in the third-person approach. For example, the idea was mentioned in Section 3 that a sensory state represents its normal distal cause, but it is unclear why we should accord a special status of representation to the causal relation in sense perception, and what it means for the distal cause to be normal, especially in the skeptical scenario. These puzzles disappear when we approach the representational relation from the first-person perspective, while taking the third-person assignment to be its extension to other people's visual experience by simulation. In other words, we construct from the first-person perspective a 3-D model of the world with a directed center, and establish the representational relation between the visual state and the facing sides of the nearest non-transparent objects in the direction of the center. We then apply the relation to other people from the third-person perspective, i.e. we ascribe to anyone with open eyes a visual experience that represents the facing sides of the nearest non-transparent objects in the direction of their eyes.[30]

2.5 Implications for Cartesian Skepticism

The previous section described the role of the representational relation in visual perception, which is in line with the correspondence theory of truth, viz. our visual state represents the facing sides of the nearest non-transparent objects in the direction of our visual sensor. They are true or false depending on the way these surfaces actually are.[31] This section examines the implication of this conception of truth for Cartesian skepticism. My focus continues to be visual experience from the first-person perspective since the role of the representational relation is clear and fundamental in it. According to Cartesian skepticism, a 3-D model of the world that is flawless for the purpose of predicting the visual experience to come may still grossly misrepresent reality because we may be BIVs manipulated by a supercomputer. This section examines whether this claim is correct, i.e. whether our visual states misrepresents reality if we are BIVs. The answer is not as straightforward as it may seem at first.

I want to begin with an analysis of visual misrepresentation in general. There are two types of visual misrepresentation. First, our 2.5-D visual experience may misrepresent the location of a surface in the 3-D world, for example, because of a reflection in a mirror.[32] Second, even if the location is correct, our visual experience may misrepresent its color, for example, because of unusual illumination. The reason for the misrepresentation of a location is easy to explain by the representational parsing of the 2-D input into a 3-D model of the world, viz. the 2-D visual input underdetermines a 3-D model. For example, with some effort the visual input to one eye (with the other eye closed) can be parsed with "a painter's eye" as representing a flat surface in front of the visual sensor. When our visual state represents a configuration of surfaces with varying depths, we may actually be receiving the visual input from a flat surface of a large *trompe-l'oeil* painting in front of us. Vision in motion makes it hard to sustain the "flat surface" interpretation. For example, in the motion parallax all visible surfaces move laterally in one direction but at varying speeds, i.e. those surfaces deeper in the 2.5-D space move more slowly while shallower surfaces (those closer to the visual sensor) move faster. This is what we should expect when we see surfaces of varying depths as the visual sensor moves laterally in the 3-D space, e.g. when we see trees at various distances in the field from a moving train. Vision in motion is hard to explain without the 3-D model with a directed center. There are no credible alternative accounts of the rich and fluid experience, and we are left with only philosophically constructed skeptical scenarios—such as the BIV hypothesis—as contenders.

The misrepresentation of the color is more complicated and goes beyond the representational parsing of the visual input. The visual input consists of the wavelengths and intensities of the lights registered on the 2-D surface of the retina. In order to focus on the non-spatial characteristics of the visual input, let us assume that the 3-D world is spatially stable and its directed center stays in the same position in the 3-D world. Our visual experience is then spatially stable, but the wavelengths and intensities of the incoming lights can still change. The complication is that not all such changes are attributed to changes in the properties of the surfaces because the incoming light is parsed further as a product of the illumination and the reflectance properties of the surface. For example, the same surface may look darker as the night falls because of the lower intensity of the light arriving in the retina, but we usually ascribe the same reflectance property to the surface. More generally, we attribute changes in the incoming light to changes in the illumination if the change is global, e.g. every surface we see suddenly looks brighter at the same time.

Change in the illumination is common as we turn the light on and off. It also occurs naturally as the sun rises in the morning, moves slowly over the course of a day, and sets in the evening. It is actually rare that the non-spatial characteristics of the visual input changes locally to signal a change in the reflectance property of the surface. The advantage of this illumination-reflectance parsing of the incoming light is similar to that of the representational parsing of the visual input. By attributing most changes in the incoming light to changes in the illumination, we maintain a mostly stable model of the reflectance properties of the surfaces, which helps us to predict the visual experience to come. The reason for misrepresentation is also similar: The incoming light underdetermines the illumination-reflectance combination. For example, we may ascribe an incorrect color (an incorrect reflectance property) to a surface under some unusual illumination, though we will realize the error when we reexamine the surface under the broad daylight.

I now turn to the skeptical scenario in which we are brains in vats whose sensory experiences are generated by the supercomputer. The answer to the question—whether the BIV's sensory experience misrepresents reality—may seem obvious. The BIV has the kind of sensory experiences we would expect in the natural world, but the objects the BIV thinks it sees only exist in the virtual reality. There is not even a visual sensor (the eye) connected to our brain, which is in a vat. It looks like the BIV's visual experience grossly misrepresents reality, even though the 3-D model of the world it constructs serves the purpose of predicting the visual experience to come perfectly well. There is, however, something peculiar in this picture.

The point of parsing the visual input by the representational relation is to improve our epistemic performance, which is to predict the visual experience to come. There is actually no problem with the BIV's prediction of the visual experience to come, and yet the representational parsing used for successful predictions is considered wrong. There is, in other words, an apparent means-end mismatch in the skeptical scenario. If the representational parsing serves its purpose perfectly well, then what is the point of evaluating the visual state negatively as grossly representing reality? One way of removing the means-end mismatch is to evaluate the visual state by the way the virtual reality is. There is no misrepresentation, in other words, provided the facing sides of the nearest non-transparent objects in the direction of the visual sensor *in the virtual reality* is the way it is represented; similarly, the visual state that must be connected to the visual sensor is the one in the virtual reality. The upshot is that our visual states do not grossly misrepresent reality in the skeptical scenario because the reality at issue is the virtual reality.[33] If the point of parsing the visual input by the representational relation is to help prediction, then it is better to evaluate the visual state by the way the virtual reality is because that is the reality that matters for the purpose of predicting the visual experience to come.

Tempting as it is, the proposal is inconsistent with the central tenet of representational parsing. Recall that the visual state plays two roles in the representational parsing: It represents a 3-D world, but it also belongs to the 3-D world that is represented. It is essential for the visual state to play these two roles. Many changes in the visual input are explained by changes in the position of the visual sensor, and this is intelligible only if the visual state that is connected to the visual sensor in the represented 3-D world is also the visual state that represents the 3-D world. The suggestion that we evaluate the accuracy of the visual state by the way the virtual reality is, instead of the way the reality of BIVs and the supercomputer is, splits the two roles by distinguishing the visual state connected to the visual sensor in the virtual reality and the visual state in the BIV that represents the virtual reality. Since the latter does not belong to the virtual reality to which the former belongs, the representational parsing of the visual input fails to explain changes in the visual input. The 3-D model constructed by the representational parsing is still successful in predicting the visual experience to come in the skeptical scenario, but for an entirely different reason, viz. the supercomputer is programed to produce the kind of visual experiences we expect to have in the natural world. In short, despite the predictive success, the BIV's visual experiences grossly misrepresent the reality they belong to.

One way of understanding the BIV's epistemic failure in the skeptical scenario is that the visual experience is implicitly self-locating.[34] Our experience of seeing a hand does not just represent the surface of a hand, but implicitly locates the visual sensor relative to the hand. Since the visual state is connected to the visual sensor in the representational parsing of the visual input, our visual experience implicitly locates the visual state as well. So, our visual experience is implicitly self-locating and the BIV's visual experience locates itself in the wrong place, i.e. in the virtual reality to which the virtual hand belongs. This is in contrast to the representation of the color, which is not self-locating even implicitly. As noted earlier, the incoming light is parsed as a product of the illumination and the reflectance property of the surface, so that the ascription of the color (the reflectance property) to the surface implicitly identifies the illumination, including the location of the illumination source. However, the illumination source needs to be located only relative to the surface, while the location of the visual state does not affect the illumination-reflectance parsing of the incoming light. Since the representation of a color is not self-locating, it makes no difference whether the visual state is housed in the brain in the skull or in the BIV.

To generalize this point beyond colors, if there is no need for locating a visual state by the representational parsing of the visual input, then it makes no difference whether the visual state is housed in the brain in the skull or in the BIV. There is therefore no Cartesian skepticism. Suppose, for example, that we parse the visual input with "a painter's eye" into a 2-D model, instead of parsing it representationally into a 3-D model with a visual state located in the model. Suppose further that the 2-D model is a model of the entire reality, so that there is no visual state in reality. The 2-D model is then the view from nowhere with no vantage point. There is no Cartesian skepticism against this model since there is no visual state to be displaced by the skeptical scenario. Cartesian skepticism arises only when we parse the sensory input representationally to construct a model that includes the sensory state located in the model.

The 2-D model with no sensory state is somewhat similar to the phenomenalist conception of reality, but the 2-D model with no sensory state still allows a distinction between true parsing and false parsing since an apparent color (a phenomenon) in the 2-D reality may be considered a product of the two factors—the "illumination" and the mostly stable color of the specific area.[35] Global changes in the apparent colors are attributed to changes in the illumination while local changes in apparent colors are attributed to changes in the colors of the specific area. Since the apparent color underdetermines the combination of the illumination and the color of the area, there can be a false ascription of a color to the area. Note, however, that the conception of truth in "true parsing" is not that of correspondence with reality since there is no sensory state that represents reality. It may be true that there is a small red circle, but it is true not because of an accurate representation. The conception of truth applicable here is that of deflationism. The upshot is that there is a distinction between the true and false parsing of the illumination-color combination even in the absence of sensory states in reality, but it is not truth in the sense of correspondence with reality. As a result, there is no Cartesian skepticism in the absence of sensory states in reality. As it stands, however, there is good reason to parse the 2-D visual input representationally into a 3-D model with visual states in it.

This chapter defended the correspondence theory of truth, according to which beliefs and statements are true if and only if they represent reality the way it actually is. The key is the representational parsing of the visual input in which the visual state plays two roles—it represents a 3-D world, but it also belongs to the 3-D world that is represented. It is these two roles of the visual state that makes truth correspondence with reality. The Cartesian skeptic challenges our belief in the natural world by a scenario that removes the visual state that represents a 3-D world from the 3-D world that is represented. It is a serious challenge because when the two roles of the visual state are split, the visual state in the BIV radically misrepresents the reality. The rest of the book takes up this challenge and examines whether our belief in the natural world is defensible.

Notes

1 Some people think that the primary bearer of truth is the *proposition*, and not the belief or the statement that expresses the proposition. The issue does not affect the discussion here as long as they accept that beliefs and statements can be true or false in the derivative sense.

2 Or a predicate represents (refers to) a set of objects if one is against the ontology of properties.

3 Some differences emerge in the understanding of correspondence when the statement or the belief is more complex. Some people understand "correspondence" as structural isomorphism, i.e. a true statement (belief) and the fact that makes it true must have identical structures. Glanzberg (2015) calls it "traditional correspondence theory" in contrast to "modern correspondence theory" with no requirement of structural isomorphism. Modern correspondence theory, which makes use of compositional semantics of the Tarskian kind, avoids some of the difficulties of traditional correspondence theory, e.g. there is no need to posit a negative fact that corresponds to a negative statement. The dispute discussed in this chapter is one between deflationism and modern correspondence theory.

4 There are obvious exceptions, e.g. the belief that there are no unicorns is true in the absence of the natural world.

5 See Davidson (1973) on the principle of charity in radical interpretation.

6 Consider a simple version of the coherence theory of truth: The belief that p is true if and only if p is coherent with other beliefs we hold. This is supposed to hold necessarily, but by combining it with the equivalence schema, we obtain the schema that p if and only if p is coherent with other beliefs we hold, whose status as necessary truth has been questioned. A similar point can be made against the pragmatic theory of truth, the epistemic theory of truth, etc. See Lewis (2001) and McGinn (2002) for more on the point.

7 This type of deflationism is sometimes called "linguistic deflationism" as distinguished from "metaphysical deflationism" (Bar-On and Simmons 2007).

8 Christopher Hill's (2002, 2014, 2016) "extended substitutionalism", which he regards as a version of deflationism, is a version of neutral deflationism in my classification. In his words, deflationists "can accommodate the traditional view that the truth condition of a thought is the state of affairs to which it semantically corresponds" (2016, p. 3175). When the content is understood representationally, the position of deflationism resembles "the identity theory of truth" (Cartwright 1987; Hornsby 1997; McDowell 1996).

9 This is the way Field (1994) characterizes deflationism.

10 See Block (1986), Field (1977), Harman (1999) and Horwich (1998) for details of conceptual role semantics.

11 We may interpret Paul Horwich's position in this way. Horwich (1990) considers the proposition to be the primary bearer of truth, leaving the impression that his version of deflationism is neutral deflationism, but Horwich (1998) also holds that the proposition expressed by a sentence is not its representational content.

12 Hill (2016) claims that the correspondence relation itself can be characterized in deflationist terms without specifying the substance of the relation. That is true in the sense that the thought x corresponds to the state of affairs y just in case for some p, x is the thought that p and y is the state of affairs that p, where the quantifier ("for some p") is substitutional (to range over expressions instead of objects). That is fine, but the characterization is of no help in clarifying the challenge of Cartesian skepticism, e.g. it does not help us determine whether it is the states of affairs in the natural world or the states of affairs in the virtual world that the BIV's thoughts should correspond to in order to be true.

13 How to understand the notion of explanation is a subject of major controversy in philosophy of science, but in the present context only empirical adequacy matters—that a psychological theory successfully explains the observed behavior in the sense of retrodicting the behavior correctly.

14 *Locus classicus* for this line of argument against intentional psychology is Stich (1983).

15 Those who suspect that "the Morning Star" and "the Evening Star" are definite descriptions, and not proper names, can replace them by "Phosphorus" and "Hesperus", respectively.

16 When our sensory state represents the world the way it actually is, it is natural to call the representation "accurate" instead of "true" because perceptual misrepresentation is a matter of degrees (one perceptual misrepresentation may be *more* inaccurate than another) while truth and falsity are considered discrete concepts with no degrees.

17 I use the term "sensory *experience*" when I refer to a sensory state from the perspective of the epistemic subject herself.

18 See Armstrong (1978, 1989, 1997) for more on internal relations.

19 I take the existential content to be one of the two standard forms of the propositional content for sensory experience, the other being the negation of existence. The reason is that we need no singular concepts such as <Gödel> or <the morning star> in capturing the content of the visual experience unless we relate it to some beliefs we already have. See Davies (1992) and McGinn (1982) for more on the existential articulation of visual experience.

20 To be more precise, it is not just a spatial location but a spatiotemporal location: In order for the sensory experience to be veridical, the property must be exemplified at the right time. My focus will be a spatial location because the right time of exemplification is almost always the time the subject is having the sensory experience, though there are some exceptions, e.g. a faraway star need not be emitting light at the moment an astronomer observes it.

21 Part of the discussion applies to other sensory modalities *mutatis mutandis*, especially to the tactile sense.

22 Some people may want to reserve the term "visual space" for the 2-D input or for the 3-D model, instead of the 2.5-D space with depths. That is only a verbal issue that does not affect the substance of the discussion, as long as the three spaces are distinguished clearly.

23 Stereo vision, which synthesizes two streams of visual input received through two eyes, is also instructive, but I will not discuss it separately since stereo vision is similar to vision in motion in many respects. It is possible to regard stereo vision as a limiting case of vision in motion in which the motion from one location (the location of one eye) to another (the location of the other eye) is instantaneous.

24 The attribution is aided by proprioception, but that is not essential. For example, we automatically attribute global shifts that take place on a movie screen to a movement of the camera without any proprioceptive input.

25 The dual nature of visual perception is emphasized, among others, by Merleau-Ponty (1962) and Gibson (1979). See also Bermúdez (1998, 2001, 2002) for some of its philosophical implications.

26 Many of these surfaces are also tilted (relative to the direction of the sensor) and warped, but I ignore these details.

27 It is possible in principle that one's visual sensor and visual state are located in distant places (cf. Dennett 1981), but the empirical evidence in neuroscience eliminates this possibility. Cartesian skepticism presents a different challenge, i.e. there may be no visual sensor (no eye) at all.

28 Some authors (Searle 1983; Siegel 2006) argue that the relation between the visual state and the perceived object is captured by our visual experience. It is certainly part of the visually informed 3-D model of the world, but the relation is not directly captured by the 2.5-D representation of the surfaces because the visual state is encased in the skull behind the visual sensor. There is, however, a clear sense in which the relation is revealed by our visual experience, viz. the 2.5-D representation locates the surfaces by their angles and distances relative to the visual sensor to which the visual state is connected.

29 There may be some concern here that the reasoning in support of representational parsing tacitly assumes that our visual perception is reliable (represents the environment accurately most of the time). If so, the epistemic justification of representational parsing may beg the question. I will address the issue of epistemic circularity in Chapter 3.

30 As mentioned earlier, once we assign representational contents to visual states, it is possible to extend it to assign representational contents to other mental states through their non-representational relations to visual states.

31 As noted earlier, perceptual misrepresentation is a matter of degrees (one perceptual misrepresentation may be *more* inaccurate than another) while truth and falsity are considered discrete concepts with no degrees. I will ignore this difference here since it does not affect the core issue. The reader may take "the true representation of reality" in the case of sense perception to mean the sufficiently accurate representation of reality.

32 The misrepresentation of the location includes errors about the angle, distance, size and shape of the surface.

33 Chalmers (2005) defends a similar view that even if we are BIVs, most of our ordinary beliefs are still true.

34 See Perry (1979) on self-locating beliefs, and Lewis (1979) on *de se* beliefs.

35 The "illumination" in the 2-D reality has no particular source location, and is uniform over the 2-D model like an ambient light.

3 The Myth of Epistemic Circularity

3.1 Epistemic Circularity

As seen in Chapter 1, meliorative social epistemology inherits the structural problems of epistemic support from traditional internalist epistemology because the basis of epistemic evaluation is restricted to the resources available to the community. This chapter takes up two of the structural problems—epistemic circularity (Sections 1–5) and epistemic regress (Sections 6–7). This is not because tackling them leads to the solution of Cartesian skepticism. Central to the challenge of Cartesian skepticism is not these two problems, but the issue of empirically equivalent alternative hypotheses. However, some presentations of Cartesian skepticism with an emphasis on the reliability of sense perception gives the impression that epistemic circularity is an important issue in Cartesian skepticism. The first five sections of the chapter show that epistemic circularity is a myth, but also that debunking the myth does not solve Cartesian skepticism.

The problem of regress, on the other hand, is not considered an issue in Cartesian skepticism. This is because the Cartesian skeptic accepts that we have a direct epistemic access to our own mental states, e.g. that I am having the visual experience of seeing a hand, or it appears to me visually that there is a hand. Reflective beliefs about the contents of our own mental states serve as the foundation of epistemic support, while the challenge is to fill the apparent gap between the reflective beliefs about the contents of our own mental states and the perceptual beliefs about the natural world in support of the latter. However, epistemologists troubled by the problem of epistemic regress may see an attempt at defending belief in the natural world as an empty exercise if it starts with the reflective beliefs as the foundation of epistemic support. The last two sections of the chapter address the issue of epistemic regress for those who question the Cartesian assumption. I propose there a variant of classic foundationalism based on the representational parsing of the visual input discussed in Chapter 2.

The problem of epistemic circularity can be understood in the following way. It is a big part of epistemology to establish the reliability of our sense perception, but any body of evidence in support of the reliability of our sense perception includes some evidence obtained by our sense perception. This is unacceptable, it is thought, because we cannot assume the reliability of sense perception when our objective is to support the reliability of sense perception. This is actually a myth: Once we give up the aspiration for absolute certainty and think of epistemic support in probabilistic terms, there is no problem of epistemic circularity.[1] To avoid distraction from the core issue, I make use of the notion of an ideal evaluator (hereafter simply "the evaluator") introduced in Chapter 1, who has an unlimited access to the best epistemic recourses—both empirical and conceptual—available to the community.

Here is an illustration of how the charge of epistemic circularity arises. Suppose the evaluator attempts to confirm the reliability of visual perception (*human* visual perception, hereafter this is assumed) by examining the physiological process of visual perception. It is not the evaluator's goal to establish the *perfect* reliability of visual perception with *absolute certainty*. So, a few errors in visual perception do not wreck her project. Nor is it necessary to eliminate all conceivable ways visual perception may be unreliable. The evaluator is only hoping that the physiological examination *raises the probability* of the hypothesis that we see things the way they actually are *in most cases*. The evidence is promising—the evaluator discovers information-preserving paths in the physiological process of visual perception as one would expect if visual perception is reliable. Let us grant here that the existence of the information-preserving paths raises the probability of the hypothesis that visual perception is generally reliable. Let us also grant, as the Cartesian skeptic does, that reflective beliefs about the contents of one's own mental states are true. So, epistemic regress is not an issue for now. For example, the evaluator is not mistaken about the content of her own visual experience.

Even with these generous assumptions, the evaluator is faced with the following charge of epistemic circularity. The evaluator examines the physiological process of visual perception in large part by her own visual perception (aided by various devices). It may be supplemented by tactile or auditory evidence, but visual evidence plays the central role. The problem is that the visual evidence has no epistemic value unless visual perception is reliable. So, when the evaluator invokes the visual evidence, she needs to make the assumption that visual perception is reliable, and the assumption makes her reasoning epistemically circular because the reliability of visual perception is the very hypothesis at issue.

This is only one illustration, but the point generalizes. Even if we assume a direct access to our own mental states, the truth of the hypothesis that human sense perception is reliable depends on the way the natural world is beyond our own mental states. For all we know about our own mental states, the natural world may be radically different from what we think it is. In order to support the hypothesis that our sense perception is reliable, we must obtain some evidence about the way the natural world is beyond our own mental states. The problem is that all such evidence ultimately comes from our sense perception. Oftentimes, the evidence comes directly from our current sense perception, for example, from the visual inspection of our environment. In some cases, we may rely on memory instead of sense perception. However, if the memory carries information on the natural world beyond our own mental states, its content must have come from sense perception at some point.[2] So, directly or indirectly we must rely on our sense perception to obtain any information about the natural world beyond our own mental states.

This point holds no matter how the argument is structured for the reliability of our sense perception, e.g. whether it is based on the track record of our sense perception, or on the physiology of our sense perception. Any such attempt is doomed—or so it seems—because sense perceptual evidence has no epistemic value unless our sense perception is reliable: When the evaluator invokes the sense perceptual evidence, she needs to make the assumption that her sense perception is reliable, and this makes her reasoning epistemically circular. The evaluator cannot avoid epistemic circularity by differentiating sensory modalities. For example, although it is not epistemically circular to support the reliability of visual perception by tactile evidence, the evaluator must then support the reliability of tactile perception. She may turn to auditory evidence, etc. but the evaluator must then support the reliability of that sense modality. The evaluator cannot keep passing the buck forever since there are only finitely many sensory modalities. At some point the evidence must come from the sensory modality whose reliability is at issue. The difficulty, in short, is that epistemic circularity seems to undercut any attempt to support the reliability of our sense perception.[3]

A similar charge of epistemic circularity arises for memory because any body of evidence in support of the reliability of memory will include some evidence that is preserved in memory.[4] Here is an illustration. Suppose the evaluator attempts to confirm the reliability of memory (*human* memory, hereafter this is assumed) from the track record. As in the earlier example, it is not the evaluator's goal to establish the *perfect* reliability of memory with *absolute certainty*. The evaluator is only hoping that the track record *raises the probability* of the hypothesis that we remember things the way they actually were *in most cases*. The evidence is promising—the evaluator notices that most memories were corroborated later by sense perception. Let us grant here that the track record of successful corroborations raises the probability of the hypothesis that memory is generally reliable. Let us also grant, as the Cartesian skeptic does, that reflective beliefs about one's own mental states are true. For example, the evaluator is not mistaken about the content of her own memory (what content her memory has). The problem, of course, is that the evaluator only *remembers* that most memories were corroborated later by sense perception. This is itself memorial evidence. So, when the evaluator introduces the track record as evidence, she needs to make the assumption that her memory is reliable. The assumption makes her reasoning epistemically circular because the reliability of memory is the very hypothesis at issue.[5]

It is sometimes suggested (e.g. BonJour 1985) that coherence among independently obtained pieces of evidence supports the hypothesis that their sources are reliable. For example, coherence of the original memory and the memory of its later corroboration seems to strengthen our general confidence in memory. An obvious challenge to this reasoning is the requirement of evidential independence. If pieces of memory are not independently obtained, but are modified—or even fabricated—in light of later experience, coherence of memories is not an indication of reliability. It is therefore crucial for the reasoning from coherence that we can establish evidential independence, but it is hard to come up with a way of establishing evidential independence without epistemic circularity, i.e. without relying on memory.[6] We also need to be careful not to conflate two questions about the role of coherence. One is whether coherence strengthens our prior confidence that is already positive.[7] However, the issue in the context of Cartesian skepticism is whether coherence can generate confidence from scratch where there is originally none. That is much harder. I do not rule out the possibility of an intricate reasoning for justification by coherence from scratch, but I do not pursue the issue in this book.[8] My solution to the problem of epistemic circularity is much simpler.

I will debunk the myth of epistemic circularity in the next four sections, but before embarking on the task, I want to clarify the scope and nature of my claim. In the two earlier examples (visual perception and memory) the problem is not that the reliability hypothesis at issue appears as a premise of the argument, but that the evidence (the visual evidence in the first example and the memorial evidence in the second example) has no epistemic value unless we assume the truth of the reliability hypothesis. This type of circularity is sometimes called *epistemic circularity* as distinguished from *logical circularity* in which the hypothesis appears as a premise of the argument, and it is the myth of *epistemic circularity* that I am going to debunk.[9] Some people suggest that epistemic circularity is not as crippling as logical circularity (e.g. Alston 1989, 1993), but most of them stop short of claiming that epistemically circular reasoning is an effective *polemical* tool.[10] For example, if you are already convinced that visual perception is reliable, then seeing information-preserving paths in the physiological process of visual perception is reassuring, but the visual evidence would not convince your opponents who question the epistemic value of the visual evidence. I am going to argue that we can use visual evidence for the reliability of visual perception in a way that is polemically effective. However, this is not because I embrace a circular reasoning. I will show instead that we can transform an epistemically circular argument into a non-circular argument that can be an effective polemical tool.

The key point in the transformation is the ambiguity of the term "assumption". There are valid forms of inference in which we "assume" the truth of some statement though its truth is not established yet. In many such cases the statement in question is never shown to be true, and in some cases it is shown to be false at the end of the reasoning. In the reasoning of *conditional proof*, for example, we prove the conditional statement of the form $p \supset q$ by "assuming" the truth of p and deriving q from the assumption. It does not matter for the purpose of conditional proof that p may be false. In the reasoning of *reductio ad absurdum*, we prove the conclusion c by "assuming" the truth of its negation $\neg c$ and deriving a contradiction from the assumption. In either of them, we do not actually accept the assumed statement as true. We make the assumption of truth only in the sense of introducing it as a hypothesis in order to find out the consequences of the assumption. This is different from making an assumption in the sense of introducing it as a *premise* of the argument or as a *presupposition* in support of a premise of the argument. When we introduce an assumption as a hypothesis, we are not committed to its truth. The charge of epistemic circularity is that the evidence used in support of the hypothesis has no epistemic value without the assumption that the hypothesis is true in the sense of the *presupposition* that it is true. I am going to show that we can always transform an epistemically circular argument into an argument in which we only make an assumption in the sense of introducing a hypothesis with no commitment to its truth.

3.2 Formal Principles of Epistemic Support

This section describes some formal principles of epistemic support that will be used in the next three sections to debunk the myth of epistemic circularity. These principles are elementary results of probability theory. So, those who are well versed in probability theory can browse through the section quickly. My intent is to explain the ideas behind these formal principles by relating them to the familiar inference rule in deductive logic, reductio ad absurdum.

Let us go over the inference rule of reductio ad absurdum. We can state it formally as follows, where $\{p_1, \ldots, p_n\}$ is the set of premises and c is the conclusion:

$$\text{If } \{p_1, \ldots, p_n, \neg c\} \vdash \bot, \text{ then } \{p_1, \ldots, p_n\} \vdash c$$

To put the idea informally, if adding $\neg c$ to $\{p_1, \ldots, p_n\}$ leads to a contradiction $\bot$, then $\neg c$ cannot be true when $p_1, \ldots, p_n$ are. So, c must be true when $p_1, \ldots, p_n$ are. It is worth noting that reductio ad absurdum is not just one rule of inference. It can be considered the basis of deductive validity in general since any valid inference in deductive logic can be transformed into an inference by reductio ad absurdum:

$$\{p_1, \ldots, p_n\} \vdash c \text{ if and only if } \{p_1, \ldots, p_n, \neg c\} \vdash \bot$$

This means that whenever we want to derive some statement c from the set of premises $\{p_1, \ldots, p_n\}$, we can add its negation $\neg c$ to $\{p_1, \ldots, p_n\}$ and derive a contradiction. The reasoning in the opposite direction (from left to right) is as follows. If $\{p_1, \ldots, p_n\} \vdash c$, then $\{p_1, \ldots, p_n, \neg c\} \vdash c$ because $\{p_1, \ldots, p_n\}$ is a subset of $\{p_1, \ldots, p_n, \neg c\}$. Meanwhile, $\{p_1, \ldots, p_n, \neg c\} \vdash \neg c$ because $\neg c$ belongs to $\{p_1, \ldots, p_n, \neg c\}$. So, if $\{p_1, \ldots, p_n\} \vdash c$, then $\{p_1, \ldots, p_n, \neg c\} \vdash c \wedge \neg c$, which is a contradiction ■

I want to make some changes in the formulation of reductio ad absurdum in anticipation of its subsequent extension to probabilistic reasoning. There are well-known degenerate cases where the argument is formally valid but the premises bear no relevance to the conclusion. This can happen in two ways, viz. the premises are inconsistent among themselves, or the conclusion is logically true. In both cases the set $\{p_1, \ldots, p_n, \neg c\}$ is inconsistent, so the argument for c from $\{p_1, \ldots, p_n\}$ is formally valid, but the conclusion c can be *anything* in the first degenerate case and the premises $\{p_1, \ldots, p_n\}$ can be *anything* in the second degenerate case. So, the premises of a deductively valid argument need not *support* the conclusion in the normal sense of the term. I want to exclude these degenerate cases from the Principle of Deductive *Support* by requiring that the set $\{p_1, \ldots, p_n\}$ be consistent, and that the conclusion c be not logically true. The latter is equivalent to the condition that $\{\neg c\}$ be consistent. The following is then the Principle of Deductive Support.

> **Principle of Deductive Support (PDS):** The set of premises $\{p_1, \ldots, p_n\}$ deductively supports the conclusion c if and only if $\{p_1, \ldots, p_n\}$ is consistent, $\{\neg c\}$ is consistent, and $\{p_1, \ldots, p_n, \neg c\}$ is inconsistent.

In other words, deductive support of c by $\{p_1, \ldots, p_n\}$ means that $\{p_1, \ldots, p_n\}$ and $\{\neg c\}$ are consistent on their own, but putting them together turn the set inconsistent.

To generalize PDS probabilistically, it is helpful to state PDS itself in probabilistic terms. First, I will assume throughout this book that the probability distribution is regular—i.e. for any statement x, $P(x) = 1$ only if x is logically true, and $P(x) = 0$ only if x is logically false. The requirement of PDS then amounts to the following: $P(p_1 \wedge \ldots \wedge p_n) > 0$, $P(\neg c) > 0$, and $P(p_1 \wedge \ldots \wedge p_n \wedge \neg c) = 0$. This is equivalent to $P(p_1 \wedge \ldots \wedge p_n) \times P(\neg c) > 0$ and $P(p_1 \wedge \ldots \wedge p_n \wedge \neg c) = 0$. Meanwhile, we can express the deductive support of c by $\{p_1, \ldots, p_n\}$ in probabilistic terms as follows: $P(c \mid p_1 \wedge \ldots \wedge p_n) = 1$ and $P(c) < 1$, where the latter excludes degenerate cases.[11] We obtain the following expression of PDS:

> **Principle of Deductive Support (PDS):** The set of premises $\{p_1, \ldots, p_n\}$ deductively supports the conclusion c, i.e. $P(c \mid p_1 \wedge \ldots \wedge p_n) = 1$ and $P(c) < 1$, if and only if $P(\neg c) \times P(p_1 \wedge \ldots \wedge p_n) > 0$ and $P(p_1 \wedge \ldots \wedge p_n \wedge \neg c) = 0$.

PDS expressed in this way follows immediately from the definition of the conditional probability. To go from left to right, suppose $P(c \mid p_1 \wedge \ldots \wedge p_n) = 1$ and

$P(c) < 1$. It follows that $P(\neg c \mid p_1 \wedge \ldots \wedge p_n) = 0$ and $P(\neg c) > 0$. It follows from $P(\neg c \mid p_1 \wedge \ldots \wedge p_n) = 0$ that $P(p_1 \wedge \ldots \wedge p_n) > 0$ and $P(p_1 \wedge \ldots \wedge p_n \wedge \neg c) = 0$. It follows from $P(\neg c) > 0$ and $P(p_1 \wedge \ldots \wedge p_n) > 0$ that $P(\neg c) \times P(p_1 \wedge \ldots \wedge p_n) > 0$ ■ To go from right to left, suppose $P(\neg c) \times P(p_1 \wedge \ldots \wedge p_n) > 0$ and $P(p_1 \wedge \ldots \wedge p_n \wedge \neg c) = 0$. It follows that $P(\neg c) > 0$, and hence $P(c) < 1$. It also follows that $P(\neg c \mid p_1 \wedge \ldots \wedge p_n) = 0$, and hence $P(c \mid p_1 \wedge \ldots \wedge p_n) = 1$■

Having stated PDS in probabilistic terms, we can now generalize it probabilistically to obtain the Principle of Probabilistic Support (PPS). When the support is probabilistic, the set of premises $\{p_1, \ldots, p_n\}$ only raises the probability of c, or $P(c) < P(c \mid p_1 \wedge \ldots \wedge p_n)$. In order for this to happen, it is not necessary that adding $\neg c$ to $\{p_1, \ldots, p_n\}$ makes the set inconsistent, or $P(p_1 \wedge \ldots \wedge p_n \wedge \neg c) = 0$. However, it is not strong enough that adding $\neg c$ to $\{p_1, \ldots, p_n\}$ lowers the joint probability of the set, or $P(p_1 \wedge \ldots \wedge p_n) > P(p_1 \wedge \ldots \wedge p_n \wedge \neg c)$. This is because this condition almost always holds except when $\{p_1, \ldots, p_n\} \vdash \neg c$, i.e. except when adding $\neg c$ to $\{p_1, \ldots, p_n\}$ makes no difference to the logical strength of the set. For example, the condition $P(p_1 \wedge \ldots \wedge p_n) > P(p_1 \wedge \ldots \wedge p_n \wedge \neg c)$ holds even if c is irrelevant to $\{p_1, \ldots, p_n\}$ and thus $P(c \mid p_1 \wedge \ldots \wedge p_n) = P(c)$.

We need a stronger condition for probabilistic support, which turns out to be the generalization of $P(p_1 \wedge \ldots \wedge p_n) \times P(\neg c) > 0$ and $P(p_1 \wedge \ldots \wedge p_n \wedge \neg c) = 0$ in PDS, viz. $P(p_1 \wedge \ldots \wedge p_n) \times P(\neg c) > P(p_1 \wedge \ldots \wedge p_n \wedge \neg c)$, which is equivalent (under the condition $P(\neg c) > 0$) to $P(p_1 \wedge \ldots \wedge p_n) > P(p_1 \wedge \ldots \wedge p_n \wedge \neg c)/P(\neg c) =_{\text{def}} P(p_1 \wedge \ldots \wedge p_n \mid \neg c)$. In other words, the condition for probabilistic support of c by $\{p_1, \ldots, p_n\}$ is that the joint probability of $\{p_1, \ldots, p_n\}$ becomes lower on the hypothetical condition that $\neg c$ is true.

> **Principle of Probabilistic Support (PPS):** The set of premises $\{p_1, \ldots, p_n\}$ probabilistically supports the conclusion c, i.e. $P(c) < P(c \mid p_1 \wedge \ldots \wedge p_n)$, if and only if $P(p_1 \wedge \ldots \wedge p_n) > P(p_1 \wedge \ldots \wedge p_n \mid \neg c)$.

We can prove the principle formally as follows. First, each side of the biconditional entails that $0 < P(p_1 \wedge \ldots \wedge p_n) < 1$ and $0 < P(c) < 1$. Under these conditions, $P(c \mid p_1 \wedge \ldots \wedge p_n) > P(c)$ iff $1 - P(\neg c \mid p_1 \wedge \ldots \wedge p_n) > 1 - P(\neg c)$ iff $P(\neg c) > P(\neg c \mid p_1 \wedge \ldots \wedge p_n)$ iff $P(p_1 \wedge \ldots \wedge p_n) \times P(\neg c) > P(p_1 \wedge \ldots \wedge p_n \wedge \neg c)$ iff $P(p_1 \wedge \ldots \wedge p_n) > P(p_1 \wedge \ldots \wedge p_n \mid \neg c)$ ■

PPS inherits a distinct feature of PDS, viz. just as the PDS allows an indirect proof of deductive validity by reductio ad absurdum, PPS allows an indirect proof of probabilistic support by *reductio ad improbabilius* (reduction to something more improbable). When it is difficult to establish directly that the premises probabilistically support the conclusion, we can establish it indirectly by showing that the negation of the conclusion reduces the joint probability of the premises. If we set aside degenerate cases, reductio ad absurdum is a limiting case of reductio ad improbabilius. If the negation of the conclusion does not just reduce the joint probability of the premises, but it reduces their joint probability to zero, then the premises do not just raise the probability of the conclusion, but they raise it to one.

So far, no distinction has been made among the members of the set $\{p_1, \ldots, p_n\}$ but some premises are new evidence, while others are old evidence or perhaps general principles assumed (presupposed) to be true. When we are interested in the impact of a particular body of evidence, it is helpful to distinguish that body of evidence e from the rest of the premises, which I will call "the background information b". So, the set of premises $\{p_1, \ldots, p_n\}$ is now written as $\{e, b\}$. I will also call the statement of interest "the hypothesis h" (instead of "the conclusion c") as is common in the literature on probabilistic confirmation. Given this terminology, the Principle of Evidential Support (PES) identifies the condition under which the body of evidence e raises the probability of the hypothesis h given the background information b, or $P(h \mid e \wedge b) > P(h \mid b)$. This is not the same as $P(h \mid e \wedge b) > P(h)$. If $P(h \mid b)$ without e is already higher than $P(h)$, then $P(h \mid e \wedge b)$ may be higher than $P(h)$ even if $P(h \mid e \wedge b)$ is not higher than $P(h \mid b)$. Since $P(h \mid e \wedge b) > P(h)$ may be due to b instead of e, the case of our interest is $P(h \mid e \wedge b) > P(h \mid b)$, where it is e that has a positive impact on h.

The Principle of Evidential Support (PES) is formally very close to PPS, except that c and $p_1 \wedge \ldots \wedge p_n$ are replaced by h and e, respectively, and the relation between h and e holds modulo b, as follows.

> **Principle of Evidential Support (PES):** A body of evidence e probabilistically supports the hypothesis h given the background information b, i.e. $P(h \mid e \wedge b) > P(h \mid b)$ if and only if $P(e \mid b) > P(e \mid \neg h \wedge b)$.

The idea remains the same as PPS, viz. PES allows an indirect proof of evidential support by *reductio ad improbabilius*. When it is difficult to establish directly that the evidence probabilistically supports the hypothesis (given the background information b), we can establish it indirectly by showing that the negation of the hypothesis reduces the probability of the evidence (given the background information b). The formal proof of PES is also similar to the proof of PPS. First, each side of the biconditional entails that $0 < P(e \mid b) < 1$ and $0 < P(h \mid b) < 1$. Under these conditions, $P(h \mid e \wedge b) > P(h \mid b)$ iff $1-P(\neg h \mid e \wedge b) > 1-P(\neg h \mid b)$ iff $P(\neg h \mid b) > P(\neg h \mid e \wedge b)$ iff $P(\neg h \mid b) \times P(e \mid b) > P(\neg h \wedge e \mid b)$ iff $P(e \mid b) > P(e \mid \neg h \wedge b)$ ■

Finally, I describe a variant of PES, which I call the Principle of Evidential Support by Contrast (PESC). PESC compares two likelihoods $P(e \mid h \wedge b)$ and $P(e \mid \neg h \wedge b)$ of h and $\neg h$, respectively, instead of $P(e \mid b)$ and $P(e \mid \neg h \wedge b)$ as in PES:[12]

> **Principle of Evidential Support by Contrast (PESC):** A body of evidence e probabilistically supports the hypothesis h given the background information b, i.e. $P(h \mid e \wedge b) > P(h \mid b)$, if and only if $P(e \mid h \wedge b) > P(e \mid \neg h \wedge b)$.

The formal proof of PESC builds on PES. First, $P(h \mid e \wedge b) > P(h \mid b)$ iff $P(e \mid b) > P(e \mid \neg h \wedge b)$ by PES. Next, $P(h \mid e \wedge b) > P(h \mid b)$ iff $P(h \wedge e \mid b) > P(h \mid b) \times P(e \mid b)$ iff $P(e \mid h \wedge b) > P(e \mid b)$. By putting them together, $P(h \mid e \wedge b) > P(h \mid b)$ iff $P(e \mid h \wedge b) > P(e \mid b) > P(e \mid \neg h \wedge b)$ iff $P(e \mid h \wedge b) > P(e \mid \neg h \wedge b)$ ∎[13] As we will see in the next section, PESC makes evaluation easier in many applications because the difference between $P(e \mid h \wedge b)$ and $P(e \mid \neg h \wedge b)$ is greater than the difference between $P(e \mid b)$ and $P(e \mid \neg h \wedge b)$.

3.3 Debunking the Myth

We are now ready to debunk the myth of epistemic circularity. As explained earlier, my target is the charge of epistemic circularity where the hypothesis is supported by the evidence that has no epistemic value unless we assume the truth of the hypothesis. This is in contrast with logical circularity where the hypothesis appears explicitly as one of the premises. For example, in the case of visual perception in Section 1, the hypothesis h_V that visual perception is reliable does not appear as a premise. The charge in the case is that the visual evidence e_V in support of the hypothesis h_V has no epistemic value unless we assume the truth of the hypothesis h_V that visual perception is reliable. The distinction may seem to be superficial because epistemic circularity is reducible to logical circularity. Take the visual evidence e_V that there are information-preserving paths in the physiological process of visual perception. We can break down e_V into two components. One of them is the *reflective* visual evidence e_{RV} that it appears to the evaluator visually that there are information-preserving paths in the physiological process of visual perception. This is the evidential component of e_V. The other component is the hypothesis h_V that visual perception is reliable. If we replace the visual evidence e_V by these two components, then h_V explicitly appears as a premise of the argument, making it a case of logical circularity.

I agree that epistemic circularity is reducible to logical circularity. However, when the hypothesis h_V is made explicit by reduction, it is accompanied by the evidential component, the reflective visual evidence e_{RV} in the example. This makes epistemic circularity a special case of logical circularity even when it is reduced to logical circularity, and the evidential component allows a transformation of a circular support into a non-circular support. Here is the general recipe for transformation. Suppose some evidence e supports the hypothesis h given the background information b, but that the evidence e has no epistemic value unless we assume the truth of the hypothesis h. When the epistemic support is epistemically circular in this way, reduce it to logical circularity by breaking down e into the hypothesis h and the evidential component e^*, i.e. replace $P(h \mid e \wedge b)$ by $P(h \mid e^* \wedge h \wedge b)$. Next, drop h from the condition $e^* \wedge h \wedge b$ of $P(h \mid e^* \wedge h \wedge b)$ to avoid logical circularity. Finally, we ask whether e^* supports h given b, i.e. whether $P(h \mid e^* \wedge b) > P(h \mid b)$, instead of asking whether e supports h given b, i.e. whether $P(h \mid e \wedge b) > P(h \mid b)$.

There is no circularity in the comparison of $P(h \mid e^* \wedge b)$ and $P(h \mid b)$ because the problematic evidence e, which is supported in part by h, is replaced by e^* and the background information b does not include the hypothesis h. However, it looks like the answer to the question whether $P(h \mid e^* \wedge b) > P(h \mid b)$ is negative—or not clearly positive—because we need e in support h, and e^* does not support e unless we assume the truth of h. The solution to this problem is the Principle of Evidential Support (PES): $P(h \mid e^* \wedge b) > P(h \mid b)$ if and only if $P(e^* \mid b) > P(e^* \mid \neg h \wedge b)$. We can prove the former indirectly by establishing the latter with the reasoning of reductio ad improbabilius, i.e. by showing that assuming the truth of $\neg h$ reduces the probability of e^*, where the assumption only means that we introduce the truth of $\neg h$ as a hypothesis. If the result is not immediately clear, we can also turn to the Principle of Evidential Support by Contrast (PESC): $P(h \mid e^* \wedge b) > P(h \mid b)$, if and only if $P(e^* \mid h \wedge b) > P(e^* \mid \neg h \wedge b)$. Again, we can prove the former indirectly by establishing the latter with the reasoning of reductio ad improbabilius, but this time by comparing the two likelihoods, $P(e^* \mid h \wedge b)$ and $P(e^* \mid \neg h \wedge b)$, instead of $P(e^* \mid b)$ and $P(e^* \mid \neg h \wedge b)$. The task is to show that assuming the truth of $\neg h$ makes the probability of e^* lower than assuming the truth of h, where the assumption only means that we introduce the truth of $\neg h$ and the truth of h, respectively, as a hypothesis. Let us see how it works with the two examples from Section 1.

In the first example, the hypothesis h_V is that visual perception is reliable, while the original visual evidence e_V is that there are information-preserving paths in the physiological process of visual perception. Since the visual evidence e_V invites the charge of epistemic circularity that it is itself supported by h_V, it is replaced by the reflective visual evidence e_{RV} that it appears to the evaluator visually that there are information-preserving paths in the physiological process of visual perception. The question after the replacement is whether e_{RV} supports h_V given the background information b, or $P(h_V \mid e_{RV} \wedge b) > P(h_V \mid b)$, where b is not dependent on visual perception.[14] This is equivalent to $P(e_{RV} \mid h_V \wedge b) > P(e_{RV} \mid \neg h_V \wedge b)$ by PESC, and the answer to the question in this form is clearly yes. First, the assumption h_V that visual perception is reliable makes e_V more probable that there are information-preserving paths in the physiological process of visual perception. This, together with the assumption h, in turn makes e_{RV} more probable, i.e. it is more probable that it appears to the evaluator visually that there are information-preserving paths in the physiological process of visual perception. In contrast, if we assume $\neg h_V$ that visual perception is *not* reliable, then there is no reason to expect e_V that there are information-preserving paths in the physiological process of visual perception, or e_{RV} that it appears to the evaluator visually that there are information-preserving paths in the physiological process of visual perception. So, assuming $\neg h_V$ makes e_{RV} less probable than does assuming h_V. We can therefore conclude by reductio ad improbabilius that the evidence e_{RV} raises the probability of the hypothesis h_V that visual perception is reliable.

For those who prefer a more formal argument, we can compare the two likelihoods $P(e_{RV} \mid h_V \wedge b)$ and $P(e_{RV} \mid \neg h_V \wedge b)$ in their expanded form, as follows:

$$\begin{aligned} P(e_{RV} \mid h_V \wedge b) &= P(e_{RV} \wedge e_V \mid h_V \wedge b) + P(e_{RV} \wedge \neg e_V \mid h_V \wedge b) \\ &= P(e_{RV} \mid e_V \wedge h_V \wedge b) \times P(e_V \mid h_V \wedge b) \\ &+ P(e_{RV} \mid \neg e_V \wedge h_V \wedge b) \times P(\neg e_V \mid h_V \wedge b) \end{aligned}$$

$$\begin{aligned} P(e_{RV} \mid \neg h_V \wedge b) &= P(e_{RV} \wedge e_V \mid \neg h_V \wedge b) + P(e_{RV} \wedge \neg e_V \mid \neg h_V \wedge b) \\ &= P(e_{RV} \mid e_V \wedge \neg h_V \wedge b) \times P(e_V \mid \neg h_V \wedge b) \\ &+ P(e_{RV} \mid \neg e_V \wedge \neg h_V \wedge b) \times P(\neg e_V \mid \neg h_V \wedge b) \end{aligned}$$

We can then match up the terms of the two expanded formulas. To start with the first term, it is clear that $P(e_{RV} \mid e_V \wedge h_V \wedge b) \times P(e_V \mid h_V \wedge b) > P(e_{RV} \mid e_V \wedge \neg h_V \wedge b) \times P(e_V \mid \neg h_V \wedge b)$ and the inequality is substantial because $\neg h_V$ makes e_V much less probable than does h_V, and $e_V \wedge \neg h_V$ makes e_{RM} much less probable than does $e_V \wedge h_V$. In contrast to the substantial inequality in the first term, the impact of the second term is negligible because both $P(e_{RV} \mid \neg e_V \wedge h_V \wedge b)$ and $P(e_{RV} \mid \neg e_V \wedge \neg h_V \wedge b)$ are close to zero, i.e. if there are no information-preserving paths in the physiological process of visual perception, then it is extremely improbable that it appears to the evaluator visually that there are information-preserving paths in the physiological process of visual perception, whether or not our visual perception is reliable. Since both $P(e_{RV} \mid \neg e_V \wedge h_V \wedge b)$ and $P(e_{RV} \mid \neg e_V \wedge \neg h_V \wedge b)$ are close to zero, the second term is negligible although $P(\neg e_V \mid \neg h_V \wedge b)$ is higher than $P(\neg e_V \mid h_V \wedge b)$. Given the substantial inequality of the first term in favor of $P(e_{RV} \mid h_V \wedge b)$ and negligible difference of the second term, $P(e_{RV} \mid h_V \wedge b)$ is more probable than $P(e_{RV} \mid \neg h_V \wedge b)$ overall.

It may be suggested with regard to the first multiplicand of the second term that under the condition of $\neg e_V$, the hypothesis $\neg h_V$ makes e_{RV} more probable than does the hypothesis h_V because unreliable visual perception makes it more probable that it appears to the evaluator that e_V is true despite the falsity of e_V. That is not incorrect. However, in the absence of information-preserving paths in the physiological process of visual perception, there are still numerous ways the physiology of the human brain can be, and only one of these numerous ways is actual under the condition of $\neg e_V$. There are therefore numerous ways other than e_V that unreliable visual perception can misrepresent the physiology of the human brain. So, under the condition of $\neg e_V$ it is still highly improbable that unreliable visual perception makes it appear to the evaluator that e_V is true. Expecting the appearance of e_V under these conditions is similar to expecting that an unreliable shooter with an unreliable gun will hit the target. In fact, it is doubtful that we can even identify and locate the brain if visual perception is unreliable. Barring extraordinary coincidence, unreliable visual perception will produce incomprehensible jumbles of visual experience.

The second example of memory and its track record can be analyzed in the same way. The hypothesis h_M is that memory is reliable, while the evaluator's original evidence e_M is that most memories were corroborated later by sense perception. The problem is that the evaluator only *remembers* that most memories were corroborated later by sense perception. Since reliance on the memorial evidence e_M invites the charge of epistemic circularity that it is itself supported by h_M, it is replaced by the reflective memorial evidence e_{RM} that it appears to the evaluator memorially that most memories were corroborated later by sense perception. The question after the replacement is whether e_{RM} supports h_M given the background information b, or $P(h_M \mid e_{RM} \wedge b) > P(h_M \mid b)$, where b is not dependent on memory. This is equivalent to $P(e_{RM} \mid h_M \wedge b) > P(e_{RM} \mid \neg h_M \wedge b)$ by PESC, and the answer to the question in this form is clearly yes. First, the assumption h_M that memory is reliable makes e_M more probable that most memories were corroborated later by sense perception. This, together with the assumption h_M, in turn, makes e_{RM} more probable that it appears to the evaluator memorially that most memories were corroborated later by sense perception. In contrast, if we assume $\neg h_M$ that memory is *not* reliable, then there is no reason to expect e_M that most memories were corroborated later by sense perception, or e_{RM} that it appears to the evaluator memorially that most memories were corroborated later by sense perception. So, assuming $\neg h_M$ makes e_{RM} less probable than does assuming h_M. We can therefore conclude by reductio ad improbabilius that the evidence e_{RM} raises the probability of the hypothesis h_M that memory is reliable.

As in the case of visual perception, we can construct a more formal argument by comparing the two likelihoods $P(e_{RM} \mid h_M \wedge b)$ and $P(e_{RM} \mid \neg h_M \wedge b)$ in their expanded form, as follows:

$$\begin{aligned} P(e_{RM} \mid h_M \wedge b) &= P(e_{RM} \wedge e_M \mid h_M \wedge b) + P(e_{RM} \wedge \neg e_M \mid h_M \wedge b) \\ &= P(e_{RM} \mid e_M \wedge h_M \wedge b) \times P(e_M \mid h_M \wedge b) \\ &+ P(e_{RM} \mid \neg e_M \wedge h_M \wedge b) \times P(\neg e_M \mid h_M \wedge b) \end{aligned}$$

$$\begin{aligned} P(e_{RM} \mid \neg h_M \wedge b) &= P(e_{RM} \wedge e_M \mid \neg h_M \wedge b) + P(e_{RM} \wedge \neg e_M \mid \neg h_M \wedge b) \\ &= P(e_{RM} \mid e_M \wedge \neg h_M \wedge b) \times P(e_M \mid \neg h_M \wedge b) \\ &+ P(e_{RM} \mid \neg e_M \wedge \neg h_M \wedge b) \times P(\neg e_M \mid \neg h_M \wedge b) \end{aligned}$$

We can match up the terms of the two expanded formulas. With regard to the first term, it is clear that $P(e_{RM} \mid e_M \wedge h_M \wedge b) \times P(e_M \mid h_M \wedge b) > P(e_{RM} \mid e_M \wedge \neg h_M \wedge b) \times P(e_M \mid \neg h_M \wedge b)$ and the inequality is substantial because $\neg h_M$ makes e_M much less probable than does h_M, and $e_M \wedge \neg h_M$ makes e_{RM} much less probable than does $e_M \wedge h_M$. In contrast to the substantial inequality in the first term, the impact of the second term is negligible because both

$P(e_{RM} \mid \neg e_M \wedge h_M \wedge b)$ and $P(e_{RM} \mid \neg e_M \wedge \neg h_M \wedge b)$ are close to zero, i.e. if it is not the case that most memories were corroborated later by sense perception, then it is extremely improbable that it appears to the evaluator memorially that most memories were corroborated later by sense perception, whether or not her memory is reliable. Since both $P(e_{RM} \mid \neg e_M \wedge h_M \wedge b)$ and $P(e_{RM} \mid \neg e_M \wedge \neg h_M \wedge b)$ are close to zero, the second term is negligible although $P(\neg e_M \mid \neg h_M \wedge b)$ is higher than $P(\neg e_M \mid h_M \wedge b)$. Given the substantial inequality of the first term in favor of $P(e_{RM} \mid h_M \wedge b)$ and negligible difference of the second term, $P(e_{RM} \mid h_M \wedge b)$ is more probable than $P(e_{RM} \mid \neg h_M \wedge b)$ overall.

3.4 The Role of the Background Information

I have described a general recipe to eliminate epistemic circularity, and applied it to two cases to show that we can support the reliability of visual perception by reflective visual evidence, and the reliability of memory by reflective memorial evidence. There are, however, some issues of the background information that need to be addressed. Take the case of visual perception, where the reliability of visual perception is supported by the reflective visual evidence e_{RV} that it appears to the evaluator visually that there are information-preserving paths in the physiological process of visual perception. The background information plays a substantial role when the evaluator interprets her visual experience in this way, e.g. what counts as an information-preserving path, how various pieces of laboratory equipment work, etc. The problem is that a big part of support for these beliefs comes from visual perception. The suspicion therefore arises that epistemic circularity may only be swept under the rug of the background information, which the evaluator assumed to be true in the sense of making the presupposition.

There is a sensible response that these beliefs are taken for granted only in the context of the current investigation, and that we can open a new investigation if any of them are challenged. There may be a charge of epistemic circularity in the new investigation, but we already have a general recipe to eliminate epistemic circularity. In other words, if there is any epistemic circularity cloaked in the background information, we can make it explicit and then eliminate it by the general recipe. Some may worry that the procedure will never remove all problematic background information. For example, in any empirical study of visual perception, we need substantial background information, some of which is supported by visual evidence. This is true when we actually conduct research, but we can always reconstruct the reasoning later to eliminate epistemic circularity.

Let us see, first, how the problem arises in formal terms. Suppose the evaluator obtains the reflective visual evidence e_{RV} that it appears to the evaluator visually that there are information-preserving paths in the physiological process of visual perception, and that e_{RV} supports the hypothesis h_V that visual perception is reliable by the reasoning of reductio ad improbabilius given the background information b^*, i.e. by deriving $P(h_V \mid e_{RV} \wedge b^*) > P(h_V \mid b^*)$ from $P(e_{RV} \mid h_V \wedge b^*) > P(e_{RV} \mid \neg h_V \wedge b^*)$. Suppose, however, that some part of the background information b^* turns out to be supported by visual evidence. The real form of comparison is then between $P(h \mid e_{RV} \wedge e_{V+} \wedge b)$ and $P(h \mid e_{V+} \wedge b)$, where e_{V+} is the additional body of visual evidence that supports that part of the background information, while b is "the clean background information" that are independent of visual evidence. We can see the problem clearly by replacing e_{V+} with $e_{RV+} \wedge h_V$ that supports e_{V+}, i.e. by comparing $P(h_V \mid e_{RV} \wedge e_{RV+} \wedge h_V \wedge b)$ and $P(h_V \mid e_{RV+} \wedge h_V \wedge b)$, where e_{RV+} is the additional body of reflective visual evidence that it appears to the evaluator visually that e_{V+}. We can see that the original evidence e_{RV} is irrelevant in the comparison because both $P(h_V \mid e_{RV} \wedge e_{RV+} \wedge h_V \wedge b)$ and $P(h_V \mid e_{RV+} \wedge h_V \wedge b)$ are trivially one. Epistemic support in this form is logically circular.

Once we make the problem clear in this way, the solution also becomes clear. First, to avoid logical circularity, we drop h_V from the conditioning part of the two conditional probabilities to compare $P(h_V \mid e_{RV} \wedge e_{RV+} \wedge b)$ and $P(h_V \mid e_{RV+} \wedge b)$, instead of comparing $P(h_V \mid e_{RV} \wedge e_{RV+} \wedge h_V \wedge b)$ and $P(h_V \mid e_{RV+} \wedge h_V \wedge b)$. There is no circularity—epistemic or logical—in the new comparison. The question is then whether e_{RV} supports h_V in the sense of $P(h_V \mid e_{RV} \wedge e_{RV+} \wedge b) > P(h_V \mid e_{RV+} \wedge b)$, which is equivalent to $P(e_{RV} \mid h_V \wedge e_{RV+} \wedge b) > P(e_{RV} \mid \neg h_V \wedge e_{RV+} \wedge b)$ by PESC. There is no a priori answer to the question. However, based on the previous judgment that $P(e_{RV} \mid h_V \wedge b^*)$ is high while $P(e_{RV} \mid \neg h_V \wedge b^*)$ is not, where b^* contains visually supported beliefs, it is reasonable to judge that $P(e_{RV} \mid h_V \wedge e_{RV+} \wedge b) > P(e_{RV} \mid \neg h_V \wedge e_{RV+} \wedge b)$. The left side is high because the conditioned part e_{RV} is the same as the previous comparison while the conditioning part $h_V \wedge e_{RV+} \wedge b$ supports the conditioning part $h_V \wedge b^*$ of $P(e_{RV} \mid h_V \wedge b^*)$, which is judged high in the previous comparison. Meanwhile, there is no reason to think $P(e_{RV} \mid \neg h_V \wedge e_{RV+} \wedge b)$ is high, i.e. if we assume that visual perception is not reliable, then there is no reason to expect e_{RV} that it appears to the evaluator visually that there are information-preserving paths in the physiological process of visual perception. Adding $e_{RV+} \wedge b$ to the conditioning part does not change it.

The reasoning above is only dependent on "the clean background information" that is independent of visual evidence. This may still seem insufficient. For example, when we turn to a body of reflective visual evidence e_{RV} in support of the reliability of visual perception h_V, the clean background information presumably includes the reliability of memory h_M because most of the evidence e_{RV} was obtained in the past and preserved in memory. Meanwhile, when we turn to a body of reflective memorial evidence e_{RM} in support of the reliability of memory h_M, the clean background information presumably includes the reliability of visual perception h_V because a large part of the evidence e_{RM} must have originally come from visual perception. The suspicion therefore arises that we are merely passing the buck back and force between the two hypotheses h_V and h_M without removing epistemic circularity in the whole scheme of epistemic support. We can extend the point to the relation between different sensory modalities, e.g. visual perception and tactile perception, but I will focus here on the relation between visual perception and memory, where the issue is clear. The solution below applies *mutatis mutandis* to the other relations.

The solution has two stages. First, I show that appropriate reflective evidence can support the hypothesis h_{VM} that *both* visual perception and memory are reliable. I will then address the suspicion that epistemic circularity is cloaked in the background information. Let e_{RVM} be the reflective visual-memorial evidence that it appears to the evaluator visual-memorially (it appears that the evaluator has the visual memory) that there are information-preserving paths in the physiological process of visual perception. As in the cases of the reliability of visual perception alone and the reliability of memory alone, e_{RVM} supports the hypothesis h_{VM} that both visual perception and memory are reliable by the reasoning of reductio ad improbabilius, i.e. by deriving $P(h_{VM} \mid e_{RVM} \wedge b) > P(h_{VM} \mid b)$ from $P(e_{RVM} \mid h_{VM} \wedge b) > P(e_{RVM} \mid \neg h_{VM} \wedge b)$ by PESC, where the reflective visual-memorial evidence e_{RVM} holds independently of the truth of h_{VM}. To put the judgment informally, assuming that both visual perception and memory are reliable makes the probability of e_{RVM} high, while there is no such expectation if we assume that either visual perception or memory is not reliable. I will not repeat the details of the argument since they are essentially the same as before.

The next step concerns the background information. Suppose much of the background information (which we now call "b^*") used in the aforementioned reasoning is supported by visual evidence, memorial evidence or visual-memorial evidence. We call this body of evidence e_{VM+}. The real form of comparison is then between $P(h_{VM} \mid e_{RVM} \wedge e_{VM+} \wedge b)$ and $P(h_{VM} \mid e_{VM+} \wedge b)$, where b is "the clean background information" that are independent of visual evidence, memorial evidence and visual-memorial evidence. It is possible that b is empty. The suspicion of epistemic circularity arises when we replace e_{VM+} with $e_{RVM+} \wedge h_{VM}$ that supports e_{VM+}, i.e. when we compare $P(h_{VM} \mid e_{RVM} \wedge e_{RVM+} \wedge h_{VM} \wedge b)$ and $P(h_{VM} \mid e_{RVM+} \wedge h_{VM} \wedge b)$, where e_{RVM+} is the body of reflective visual evidence, reflective memorial evidence, and reflective visual-memorial evidence, which in combination with h_{VM} supports e_{VM+}. We can see that the evidence e_{RVM} is irrelevant in the comparison because both $P(h_{VM} \mid e_{RVM} \wedge e_{RVM+} \wedge h_{VM} \wedge b)$ and $P(h_{VM} \mid e_{RVM+} \wedge h_{VM} \wedge b)$ are trivially one. Epistemic support in this form is logically circular.

The solution is the same as before. First, to avoid logical circularity, we drop h_{VM} from the conditioning part of the two conditional probabilities, to compare $P(h_{VM} \mid e_{RVM} \wedge e_{RVM+} \wedge b)$ and $P(h_{VM} \mid e_{RVM+} \wedge b)$, instead of comparing $P(h_{VM} \mid e_{RVM} \wedge e_{RVM+} \wedge h_{VM} \wedge b)$ and $P(h_{VM} \mid e_{RVM+} \wedge h_{VM} \wedge b)$. There is no circularity—epistemic or logical—in the new comparison. The question is then whether e_{RVM} supports h_{VM} in the sense of $P(h_{VM} \mid e_{RVM} \wedge e_{RVM+} \wedge b) > P(h_{VM} \mid e_{RVM+} \wedge b)$, which is equivalent to $P(e_{RVM} \mid h_{VM} \wedge e_{RVM+} \wedge b) > P(e_{RVM} \mid \neg h_{VM} \wedge e_{RVM+} \wedge b)$ by PESC. There is no a priori answer to the question. However, based on the previous judgment that $P(e_{RVM} \mid h_{VM} \wedge b^*)$ is high while $P(e_{RVM} \mid \neg h_{VM} \wedge b^*)$ is not, it is reasonable to judge that $P(e_{RVM} \mid h_{VM} \wedge e_{RVM+} \wedge b) > P(e_{RVM} \mid \neg h_{VM} \wedge e_{RVM+} \wedge b)$. The left side is high because the conditioned part e_{RVM} is the same as the previous comparison while the conditioning part $h_{VM} \wedge e_{RVM+} \wedge b$ supports the conditioning part $h_{VM} \wedge b^*$ of $P(e_{RVM} \mid h_{VM} \wedge b^*)$, which is judged high in the previous comparison. Meanwhile, there is no reason to think that $P(e_{RVM} \mid \neg h_{VM} \wedge e_{RVM+} \wedge b)$ is high, i.e. if we assume that visual perception or memory is not reliable, then there is no reason to expect e_{RVM} that it appears to the evaluator visual-memorially that there are information-preserving paths in the physiological process of visual perception. Adding $e_{RVM+} \wedge b$ to the conditioning part does not change it.

To summarize, the general recipe to eliminate epistemic circularity by the reasoning of reductio ad improbabilius is applicable to the cases of the reliability of visual perception, the reliability of memory, and the reliability of visual perception and memory. We can extend the application to other cases in similar ways. The recipe is effective even when epistemic circularity is initially cloaked in the background information, or dispersed in the whole scheme of epistemic support. Once we conceive of epistemic support in probabilistic terms, there is no epistemic circularity that we cannot eliminate by the general recipe.

3.5 Empirically Equivalent Hypotheses

I have shown that we can transform an epistemically circular argument into an argument with no circularity. This undermines the common view that epistemically circular argument is not an effective polemical tool, viz. though an epistemically circular argument itself is not an effective polemical tool, we can make it polemically effective by eliminating epistemic circularity. There is, however, a different kind of worry about epistemic circularity: Embracing epistemic circularity legitimizes many arguments that are apparently unacceptable. A frequently cited example is a self-certifying testimony.[15] Suppose you receive a phone call from a telemarketer, whom you suspect is a scam artist. So, you ask the telemarketer whether she is telling the truth, and the telemarketer emphatically answers that she is. The common reaction to the example is that her statement provides no epistemic support for the hypothesis that she is honest because the statement has no epistemic value unless we assume that she is honest. The argument is epistemically circular.

To make my stance clear, I do not embrace epistemic circularity but am only proposing a general recipe for its elimination. The worry, however, is that the recipe legitimizes many apparently unacceptable arguments by eliminating epistemic circularity from them. As I stressed in Chapter 1, meliorative epistemology does not try to accommodate pre-theoretical judgments, but to guide them. So, there is no need to revise the theory even if some of its consequences are counterintuitive. After all the general recipe is supported by PESC, which is a theorem of the probability calculus. We cannot restrict the application of a mathematical theorem just because we don't like the result. However, cases like self-certifying testimony makes it worthwhile to investigate them to see why the reasoning seems unacceptable. The analysis below reveals that the telemarketer's testimony fails to support the hypothesis of her honesty—but the failure has nothing to do with epistemic circularity. More importantly, the analysis also reveals that the elimination of epistemic circularity by the general recipe does not solve Cartesian skepticism. The central issue in Cartesian skepticism is not epistemic circularity.

Let us see the charge of epistemic circularity in the telemarketer case, and how the general recipe works. The hypothesis h in the case is that the telemarketer is honest, while the apparent evidence e is that she is telling the truth. The apparent evidence e supports h, or $P(h \mid e \wedge b) > P(h \mid b)$, but the actual evidence you have is the testimonial evidence e_T that the telemarketer says she is telling the truth. Since the apparent evidence e is supported by the testimonial evidence e_T and the hypothesis h to be included in the background information b^*, the argument is epistemically circular. We can make circularity explicit, by replacing e by $e_T \wedge h$ to compare $P(h \mid e_T \wedge h \wedge b)$ and $P(h \mid h \wedge b)$, where b is the clean background information without h. Since both of them are trivially one, there is no confirmation of h by e_T. As in other cases, we can eliminate circularity by applying the general recipe, i.e. by dropping h in the conditioning part of $P(h \mid e_T \wedge h \wedge b)$ and $P(h \mid h \wedge b)$, to compare $P(h \mid e_T \wedge b)$ and $P(h \mid b)$, instead of comparing $P(h \mid e_T \wedge h \wedge b)$ and $P(h \mid h \wedge b)$. The question is then whether $P(h \mid e_T \wedge b) > P(h \mid b)$, which is equivalent to $P(e_T \mid h \wedge b) > P(e_T \mid \neg h \wedge b)$ by PESC. Everything is the same up to this point.

The difference is that the inequality does not hold in this case, viz. given the common background information about the telemarketer, you fully expect that she will say she is telling the truth even if she is not honest. Of course, you also expect that she will say she is telling the truth if she is honest. As a result, $P(e_T \mid h \wedge b) = P(e_T \mid \neg h \wedge b)$, which is equivalent to $P(h \mid e_T \wedge b) = P(h \mid b)$. The testimonial evidence e_T does not support the hypothesis h. The failure of the argument has nothing to do with circularity—logical or epistemic. The argument fails for a simple reason: If the truth and falsity of the hypothesis makes no difference in the probability of the observation, then the observation does not support the hypothesis. In the present case whether the telemarketer is honest or not makes no difference in the probability of the testimonial evidence e_T that the telemarketer says she is telling the truth. The reasoning of reductio ad improbabilius does not get off the ground in the absence of inequality in the likelihoods.

It is helpful to compare the telemarketer case with another case that is similar but has a different result. Suppose you are impressed by the teachings of a guru, who is believed by his followers to be infallible. You are not totally sold and ask the guru whether he is infallible. The guru answers that he is indeed infallible. The common reaction is the same as in the telemarketer case, viz. the guru's answer does not support the hypothesis of his infallibility because you cannot trust his words unless you already accept that he is infallible. But the guru case is different and his statement does support the hypothesis given the sensible background information. To see why, compare the two likelihoods $P(e_T \mid h \wedge b)$ and $P(e_T \mid \neg h \wedge b)$, where h is the hypothesis that the guru is infallible, while e_T is the testimonial evidence that the guru says he is infallible. Let us assume that the background information is not definitive about the guru's infallibility, and the guru could have admitted that he is fallible in response to your question. Let us also assume, for the sake of simplicity, that the guru gives a straight answer, yes or no, to your question. Given this setting, $P(e_T \mid h \wedge b) = 1$ because e_T is the only answer the guru will give if he is infallible. Meanwhile, $P(e_T \mid \neg h \wedge b) < 1$ because it is possible that the guru admits that he is fallible on condition that he is fallible. By putting them together, we obtain $P(e_T \mid h \wedge b) > P(e_T \mid \neg h \wedge b)$, and it follows from this by the reasoning of reductio ad improbabilius that $P(h \mid e_T \wedge b) > P(h \mid b)$. The guru's answer therefore supports the hypothesis of his infallibility. It should be noted that the rise in the probability will be small because it is highly improbable that anyone is infallible, but if the hypothesis has even a miniscule chance of being true, then the testimonial evidence e_T raises the probability to some degree. The difference between the two cases is that the dishonest telemarketer will not admit that she is deceptive, while the fallible guru may admit his fallibility.

In sum, even in the case of an apparently unacceptable argument, the general recipe eliminates epistemic circularity, but in some of them the evidence e does not support the hypothesis h because reductio ad improbabilius is not applicable, i.e. assuming the truth of $\neg h$ does not make the probability of e lower than does assuming the truth of h. However, in others the evidence supports the hypothesis despite the contrary appearance. The key is whether there is any difference between the two likelihoods, $P(e \mid h \wedge b)$ and $P(e \mid \neg h \wedge b)$. It has nothing to do with epistemic circularity.

An important implication of this observation is that epistemic circularity is not an important issue in Cartesian skepticism. We can remove epistemic circularity in any argument for the natural world hypothesis, e.g. we can cite reflective visual evidence in support of the hypothesis that visual perception is reliable with no circularity—logical or epistemic. However, this does not solve Cartesian skepticism, where the issue is not whether some evidence supports the natural world hypothesis—the answer is clearly yes—but whether some evidence favors the natural world hypothesis specifically over the deceiver hypothesis, such as the BIV hypothesis, which is empirically equivalent to the natural world hypothesis. To make the point clear, it is helpful to formulize the concept of empirical equivalence.

The traditional conception of empirical equivalence is deductive, viz. two hypotheses h_1 and h_2 are empirically equivalent given the background information b just in case for any possible observation e, h_1 logically entails e if and only if h_2 logically entails e, i.e. $\forall e(\{h_1, b\} \vdash e \equiv \{h_2, b\} \vdash e)$. However, there is also a Bayesian formalization of empirical equivalence, according to which two hypotheses h_1 and h_2 are empirically equivalent given the background information b just in case for any possible observation e, its conditional probabilities given h_1 and given h_2 are the same on condition of b, i.e. $\forall e(P(e \mid h_1 \wedge b) = P(e \mid h_2 \wedge b))$.[16] The Bayesian requirement for empirical equivalence is much stronger than the traditional requirement. It is quite common that although neither h_1 nor h_2 logically entails e or its negation given b, one of them, say h_1, makes some observation e more probable than does h_2 given b, or $P(e \mid h_1 \wedge b) > P(e \mid h_2 \wedge b)$. In such cases h_1 and h_2 may still be empirically equivalent given b in the traditional deductive sense, but not in the Bayesian sense. Since the evidence e in such cases can tip the balance between the two hypotheses (dramatically in some cases), it is more in line with the spirit and the role of empirically equivalent hypotheses in Cartesian skepticism that we not regard them as empirically equivalent.[17] I will therefore adopt the Bayesian conception of empirical equivalence, according to which h_1 and h_2 are not empirically equivalent if their likelihoods are different. In what follows, empirical equivalence means empirical equivalence in the Bayesian sense that $P(e \mid h_1 \wedge b) = P(e \mid h_2 \wedge b)$ for any possible observation e.

Returning to Cartesian skepticism, the issue is whether some evidence e favors the natural world hypothesis h_{NW} over the BIV hypothesis h_{BIV}, where e is a possible body of reflective evidence. We can see immediately that the answer is no since h_{NW} and h_{BIV} are empirically equivalent given the empty background information, i.e. $P(e \mid h_{NW}) = P(e \mid h_{BIV})$ for any possible evidence e. In Cartesian skepticism we focus on the two hypotheses h_{NW} and h_{BIV} in disregard of other hypotheses, which amounts to evaluating the impact of evidence e on h_{NW} on condition of the disjunction $h_{NW} \vee h_{BIV}$, i.e. to compare $P(h_{NW} \mid e \wedge (h_{NW} \vee h_{BIV}))$ and $P(h_{NW} \mid h_{NW} \vee h_{BIV})$. Since h_{NW} and h_{BIV} are mutually exclusive, it follows from $P(e \mid h_{NW}) = P(e \mid h_{BIV})$ that:

$$
\begin{aligned}
P(e \mid h_{NW} \wedge (h_{NW} \vee h_{BIV})) &= P(e \mid h_{NW}) \\
&= P(e \mid h_{BIV}) \\
&= P(e \mid h_{BIV} \wedge (h_{NW} \vee h_{BIV})) \\
&= P(e \mid \neg h_{NW} \wedge (h_{NW} \vee h_{BIV}))
\end{aligned}
$$

We conclude by PESC that $P(h_{NW} \mid e \wedge (h_{NW} \vee h_{BIV})) = P(h_{NW} \mid h_{NW} \vee h_{BIV})$. In other words, if h_{NW} and h_{BIV} are the only contenders, then no evidence supports the natural world hypothesis.

Of course, if we consider other contenders that are not empirically equivalent to h_{NW}, then some evidence e supports h_{NW} in the sense that $P(h_{NW} \mid e) > P(h_{NW})$. However, any such evidence e also supports h_{BIV} and never favors h_{NW} over h_{BIV} in that the ratio of the two probabilities remains the same, as follows:

$$\frac{P(h_{NW} \mid e)}{P(h_{BIV} \mid e)} = \frac{P(h_{NW} \wedge e) / P(e)}{P(h_{BIV} \wedge e) / P(e)} = \frac{P(h_{NW} \wedge e)}{P(h_{BIV} \wedge e)} = \frac{P(e \mid h_{NW})P(h_{NW})}{P(e \mid h_{BIV})P(h_{BIV})}.$$

It follows from this and $P(e \mid h_{NW}) = P(e \mid h_{BIV})$ that:

$$\frac{P(h_{NW} \mid e)}{P(h_{BIV} \mid e)} = \frac{P(h_{NW})}{P(h_{BIV})}.$$

So, no possible evidence favors the natural world hypothesis over the BIV hypothesis, and this has nothing to do with epistemic circularity. The central issue of Cartesian skepticism is empirically equivalent alternative hypotheses, and not epistemic circularity.

3.6 The Problem of Epistemic Regress in the Bayesian Setting

This section addresses the problem of epistemic regress. As mentioned already, it is not necessary to address this problem in the dialectic of Cartesian skepticism because the Cartesian skeptic accepts that we have a direct epistemic access to our own mental states, and beliefs about our own mental states serve as the foundation of epistemic support. I discuss the problem, nevertheless, for those who find this form of foundationalism questionable. First, I formulate the problem of epistemic regress (Section 6) because the problem in the Bayesian setting is sometimes misunderstood. I will then propose a solution to the problem (Section 7) that draws on the analysis of truth in Chapter 2.

To recall the problem of epistemic regress, the evaluator may find some statement in good epistemic standing on the basis of some other statement she already finds in good epistemic standing, and she may have found the latter in good epistemic standing on the basis of still other statement she had already found in good epistemic standing, and so on. The problem of regress arises about the chain of epistemic support as to whether it comes to an end at some foundational level. If the answer is yes, then three questions need to be answered: What is the foundation of epistemic support; how it acquires its status as the foundation; and how it supports other beliefs. I want to begin with the third question.

In the classic setting where the goal of cognition is to attain absolute certainty, epistemic support must be deductive entailment. We have good reason to accept a statement if it is deductively entailed by a statement we have good reason to accept, and we have good reason to accept the latter if it is deductively entailed by a statement we have good reason to accept. It is natural in this setting to ask where the chain of deductive entailment begins, or what the first statements are that we have good reason to accept with no support from other statements we accept. The underlying assumption is that we cannot use a statement in support of other statements unless we have good reason to accept it. This is a sensible assumption in the classic setting, but not in the Bayesian setting where epistemic support need not be deductive entailment. This is easy to see by an example.

Suppose the conditional probability $P(h_1 \mid h_2 \wedge b)$ of h_1 on condition of h_2 is sufficiently high for the acceptance of h_1. For example, $P(h_1 \mid h_2 \wedge b) = 0.95$ while the threshold of acceptance t is 0.9, so that $P(h_1 \mid h_2 \wedge b) \geq t$.[18] Of course, $P(h_1 \mid b)$ could still fall short of the threshold since $P(h_1 \mid h_2 \wedge b)$ is conditional on h_2. So, it is natural to ask whether the probability $P(h_2 \mid b)$ of h_2 that supports h_1 is sufficiently high in the sense of $P(h_2 \mid b) \geq t = 0.9$. However, it turns out that $P(h_2 \mid b) \geq t$ is neither necessary nor sufficient for making the probability of h_1 sufficiently high. First, there are many cases where $P(h_1 \mid b)$ is sufficiently high even though $P(h_2 \mid b)$ is not sufficiently high, e.g. $P(h_2 \mid b) = 0.8 < t$ while $P(\neg h_2 \mid b) = 0.2$ and $P(h_1 \mid \neg h_2 \wedge b) = 0.85$, as follows:

$$\begin{aligned} P(h_1 \mid b) &= P(h_1 \wedge h_2 \mid b) + P(h_1 \wedge \neg h_2 \mid b) \\ &= P(h_1 \mid h_2 \wedge b) \times P\left(h_2 \mid b\right) + P(h_1 \mid \neg h_2 \wedge b) \times P\left(\neg h_2 \mid b\right) \\ &= 0.95 \times 0.8 + 0.85 \times 0.2 \\ &= 0.93 \\ &\geq 0.9 \end{aligned}$$

Meanwhile, there are also many cases where $P(h_1 \mid b)$ is not sufficiently high even though $P(h_2 \mid b)$ is sufficiently high, e.g. $P(h_2 \mid b) = 0.9 \geq t$ while $P(\neg h_2 \mid b) = 0.1$ and $P(h_1 \mid \neg h_2 \wedge b) = 0.2$, as follows:

$$\begin{aligned} P(h_1 \mid b) &= P(h_1 \wedge h_2 \mid b) + P(h_1 \wedge \neg h_2 \mid b) \\ &= P(h_1 \mid h_2 \wedge b) \times P(h_2 \mid b) + P(h_1 \mid \neg h_2 \wedge b) \times P(\neg h_2 \mid b) \\ &= 0.95 \times 0.9 + 0.2 \times 0.1 \\ &= 0.875 \\ &< 0.9 \end{aligned}$$

In other words, though it is tempting to ask whether the probability $P(h_2 \mid b)$ is sufficiently high when we learn that the conditional probability $P(h_1 \mid h_2 \wedge b)$ is sufficiently high, that is not the right question to ask in the Bayesian setting.

In fact the problem of regress in the familiar form does not arise in the Bayesian setting. In order for some statement to support some other statement, it is neither necessary nor sufficient for the supporting statement to be supported in the sense of having a sufficiently high probability. Since there is no need for each statement in a chain of epistemic support to have a sufficiently high probability, it is not necessary to find a special category of statements that serve the foundation of epistemic support.[19] However, there is a problem of regress in a different form with regard to *changes* in the probability. It is common that we come to accept a statement we did not accept previously because we change the probability we assign to the statement. The change in the probability may be triggered by a change in the probability of some other statement that is related to it. This change may in turn be triggered by a change in the probability of some other statement that is related to it. The chain of changes in probability must have a beginning. What started the series of changes in probability?

If we confine ourselves to the empirical source, the obvious answer is the sensory input we receive. We change the probability of some statement when we see, hear, touch, smell or taste something. However, the question of interest here is not the causal origin of the change, but the epistemic ground for the change, which must be some conscious state. The most plausible candidate is therefore our sensory experience such as the experience of seeing a hand, which gives us reason for changing probabilities we assign to relevant statements. There is, however, a well-known ambiguity between the state and the content with regard to sensory experience.[20] If sensory experience is the epistemic ground for changing probabilities, is it the content of the experience or the presence of the sensory state with that content that initially triggers the change in probabilities? Depending on which way we interpret the suggestion, we encounter different problems.

To make the difference between the two interpretations easy to see, I adopt the notation ⌜<*L*, *P*>⌝ that refers to the representational state with the content that *P* in the language *L*. For example, when the language is French and the content is that here is a hand, ⌜<Fr, that here is a hand>⌝ refers to the French sentence “Voici une main”. Monolingual English speakers cannot use the sentence because it is not an English expression, but they can mention it. The same is true when *L* is the subject’s visual language V_S. For example, ⌜<V_S, that here is a hand>⌝ refers to the S’s visual state with the content that here is a hand. We cannot use it in English because ⌜<V_S, that here is a hand>⌝ is not an English expression, but we can mention it. It is some state in S’s mind (S’s brain).

Of course, S herself can use ⌜<V_S, that here is a hand>⌝—it can be present in her mind. So, one interpretation of the suggestion above is that the sensory state as it is used by the subject S is the epistemic ground for changing the probability that S assigns to its content. For example, the visual state ⌜<V_S, that here is a hand>⌝ as it is used by S is the ground for S to change the probability to assign to the statement that here is a hand. This is psychologically natural from S's perspective because *that here is a hand* is the content of her visual experience. The problem, however, is the possibility of misrepresentation: There may not be a hand in front of S when she has the visual experience of seeing a hand. Further, there is no clear epistemic reason—unless we make some additional assumptions—for changing the probability of the statement in light of the visual experience. In other words, though it is psychologically natural for the subject to change the probability, its epistemic legitimacy is suspect.

The other interpretation seems more promising, viz. it is not the content of the sensory state, but the presence of the sensory state with the content, that is the epistemic ground for S to change the probability that the sensory state is present. For example, the presence of the visual state ⌜<V_S, that here is a hand>⌝ is the epistemic ground for changing the probability that the visual state with the content that there is a hand is present in her mind. Unlike the first interpretation, the possibility of visual misrepresentation is not a problem because the statement at issue is not that here is a hand, but the presence of the sensory state with that content. The problem, however, is that the presence of the sensory state is not the content of the visual state ⌜<V_S, that here is a hand>⌝ as it is used by S. Its content is simply *that here is a hand*. So, S must have another mental state, a reflective belief whose content is the presence of that sensory state in her mind. But how can S form the reflective belief?

One account frequently discussed is introspection, viz. we have an introspective access to our own mental state through an inner sense, comparable to the extrospective access to the external world through vision, touch, etc. The idea is that when there is a sensory state in our mind, the introspective access allows us to form a distinctive mental state whose content is the presence of the sensory state. The obvious problem is again the possibility of misrepresentation. If introspection is comparable to extrospection, it seems possible that we falsely believe that a certain sensory state is present in our mind just as it is possible that we falsely believe a certain object is present in the external world. There is no clear epistemic reason to think that introspection is immune to an error. Further, there is no clear epistemic reason—unless we make some additional assumptions—for changing the probability we assign to the presence of a sensory state in light of the (possibly false) inner perception.

In addition, the notion of introspection is mysterious when we distinguish the state and the content of the sensory state. Do we have an introspective access to the sensory state itself (some state of the brain if one accepts physicalism) or to its content (e.g. that here is a hand)? If the answer is the former, then there is a further challenge of identifying the content of the sensory state. Even if we have an epistemic access to some representational state (e.g. the French sentence "Voici une main", some state of the brain, etc.), we may not be able to tell its content. Meanwhile, if the answer to the question is the latter, the inner sense must allow us to perceive abstract entities, e.g. the content that here is a hand, which is neither the sensory state nor a hand in the external world. Though the possibility cannot be ruled out, it is a highly unusual epistemic capability and there is no compelling reason to embrace the suggestion—especially because a more sensible alternative is available.

3.7 Ascent to Reflective Belief

The alternative account accepts that we form a mental state whose content is the presence of a sensory state, but denies that we have an introspective access to the sensory state or to its content. Instead, we form a reflective mental state by deriving it from the sensory state in question.[21] The sensory state itself does not represent our own mental state or its content. However, it is a conscious state and the subject is aware of its content, e.g. that here is a hand, as the way her environment is. The subject can therefore form a mental state whose content is the presence of the sensory state by reinterpreting the content as the content of her sensory experience instead of the way her environment is. To put it formally, given the sensory state $\ulcorner\langle V_S$, that here is a hand$\rangle\urcorner$ S forms the reflective belief $\ulcorner\langle M_S, (\exists x)(x \in \langle V_S$, that here is a hand$\rangle)\rangle\urcorner$, where M_S is S's mental language that includes her visual language V_S. The move is a semantic ascent from a sensory state as it is used by S to a reflective belief that refers to it. It is similar to the move from "Snow is white" to " 'Snow is white' is true" which is also a semantic ascent from a sentence as it is used by the speaker to a meta-linguistic sentence that refers to the original sentence. I call them, respectively, "reflective semantic ascent" and "alethic semantic ascent".

The account by reflective semantic ascent is sensible but there is one problem to overcome, and we can see the problem from the difference between reflective semantic ascent and alethic semantic ascent. In alethic semantic ascent, the derived sentence is logically equivalent to the original sentence, e.g. snow is white if and only if "Snow is white" is true. That is not the case in reflective semantic ascent, e.g. it is possible that the original visual state that here is a hand falsely represents the environment, while the derived reflective belief that the visual state that here is a hand is present (to put it informally, that it appears to me visually that here is a hand) is true. In fact, the truth of the derived reflective belief is not affected by the truth and falsity of the original visual state. So, reflective semantic ascent is not a form of logical inference. This is important because if reflective semantic ascent is a form of logical inference, we must ascertain the truth of the original visual state first.[22] What makes reflective semantic ascent special is that the derived reflective belief is true whether or not the original sensory state is true. The crucial question for this account is, therefore, what makes reflective semantic ascent a legitimate derivation.

The answer to the question is found in the analysis of truth put forward in the previous chapter, viz. it is the representational parsing of the sensory input that underwrites the ascent from the sensory state to the reflective belief about its content. In the case of vision, for example, the 2-D visual input is parsed into a 3-D world with a directed center, with the tacit understanding that the visual state belongs to the 3-D world, receives information from its environment, and represents the facing sides of the nearest non-transparent objects relative to the directed center (the eye). On this understanding, seeing an object in a certain way means that the visual state that represents the object in that way is present in the 3-D world. The epistemic subject S can therefore derive the reflective belief ⌜<M_S, $(\exists x)(x \in$ <V_S, that here is a hand>)>⌝ from the visual state ⌜<V_S, that here is a hand>⌝ by reflective semantic ascent.

The resulting view is a version of classic foundationalism where basic beliefs that trigger changes in probabilities are reflective beliefs about the contents of one's own sensory states. Those beliefs are true not because we can observe our own mental states by inner sense, but because the contents of the sensory states used by the subject are recycled as the contents of the sensory states mentioned in her reflective beliefs. Without any additional empirical input, the subject simply adds the mental prefix ⌜<M_S, that I have the visual experience>⌝ to the visual state ⌜<V_S, that here is a hand>⌝ to form the reflective belief ⌜<M_S, that I have the visual experience that here is a hand>⌝, or that it appears to me visually that here is a hand, to put it informally.[23] The ultimate epistemic grounds (what are originally given to us) are therefore the sensory states. The sensory states may misrepresent the environment, but that does not affect the legitimacy of reflective semantic ascent. The reflective beliefs about their contents (what contents the sensory states have) are true even if the contents do not match the way the environment actually is.

So, it is the representational parsing of the sensory input that legitimizes reflective semantic ascent, and as seen in the previous chapter, the representational parsing of the sensory input is supported by the kind of 2-D visual input we receive. For example, a complex but orderly sequence of visual experience in vision in motion matches the prediction based on the hypothesis that our visual states represent the facing sides of the nearest non-transparent objects relative to the directed center. It may be suggested that we should step back and start from *qualia* with no representational contents since the raw visual input has no representational structure. That is not necessary. If we start with the hypothesis of the sensory representation of the environment and obtain representational sensory experience of the kind predicted by the hypothesis, which we would not expect otherwise, then the hypothesis is supported by the sensory experience.

Though it is not necessary to start from qualia, I do not object to such an approach. There is a proposal, for example, that the subject is *acquainted* with the phenomenal qualities of her own sensory states, where acquaintance is direct awareness that is not mediated by inner sense. In the state of acquaintance, we are directly aware—it is suggested—of felt qualities of pain, coldness, redness, etc. The key point of the account is that these phenomenal qualities are intrinsic properties of one's sensory states, and not their representational contents.[24] This is both a strength and a weakness of the theory. The strength is that the acquaintance theory is consistent with the thesis of mind-body supervenience, i.e. the subject's mental properties are supervenient on the intrinsic physical properties of her brain, so that two subjects are different in their mental properties only if there is some difference in the intrinsic physical properties of their brains. The weakness is the phenomenological strain. As is often pointed out, we focus our attention toward the external world when we answer the questions about one's own mental states, e.g. that I see a hand in front of me, that I believe there will be a third world war, etc.[25] We do not attend to the intrinsic properties of our mind, as the acquaintance theory suggests. Even in the case of pain we locate it in some part of our body instead of our brain (unless it is a headache) where the sensory state exists.

In my view neither point is decisive in choosing between the qualia approach and the representational approach. As explained in the previous chapter, the representational parsing of the sensory input is consistent with mind-body supervenience. Phenomenology is not decisive either. Although the representational parsing of the sensory input seems to be automatic in our everyday life, we may suspend the everyday assumption and focus on the qualities of our sensory experience. So, I am open to the idea of starting from the intrinsic qualities of the sensory experience. It is also worth noting that one type of acquaintance theory (Gertler 2001; Chalmers 2003) resembles that of reflective semantic ascent proposed in this section, viz. when the subject becomes acquainted with the phenomenal quality of her sensory state, the subject's state of being acquainted with it comprises the sensory state as its part. In other worlds, the sensory state is recycled in the state of acquaintance, just as the sensory state is recycled in the reflective belief about its content in the ascent account. My main concern, however, is that there is no clear path from the state of acquaintance to the belief in the natural world. If we are only acquainted with the intrinsic properties of the sensory experience, we need some additional theoretical reason for assigning representational contents to the sensory states. There may be a sensible way of developing such a theory, but I suspect the resulting account will be similar to the representational parsing of the sensory input. So, my preference is to assume the representational structure from the outset in the sense of introducing it as a hypothesis.

To summarize, the chapter proposed answers to the two common worries about the structure of epistemic support. The problem of epistemic circularity is a myth because there is a general recipe to eliminate it so that we can apply the reasoning of reductio ad improbabilius. There is no problem of epistemic regress either. First, with regard to probabilistic support, there is no regress of support that motivates foundationalism in the classic setting where epistemic support must be deductive entailment. Second, with regard to epistemic grounds that trigger changes in probabilities, reflective beliefs about the contents of one's sensory states serve as the basic beliefs. These reflective beliefs are obtained by reflective semantic ascent from the sensory states. These answers to the two structural problems of epistemic support do not solve Cartesian skepticism because the skeptic is proposing an alternative hypothesis that is empirically equivalent to the natural world hypothesis, and the reasoning of reductio ad improbabilius is ineffective when the two hypotheses are empirically equivalent. The next two chapters develop a new Bayesian approach to Cartesian skepticism to address the problem of empirically equivalent alternative hypotheses.

Notes

1 See Shogenji (2000, 2006, 2013) for previous attempts at solving the problem of epistemic circularity along the same line.

2 This is also true of testimonial evidence, viz. we may rely on testimony instead of sense perception or memory, but if testimony carries information on the natural world beyond our mental states, its content must have originally come from sense perception of someone in the epistemic community at some point. Testimonial evidence is set aside here because the notion of the ideal evaluator with an unlimited access to the epistemic resources of the community makes testimonial evidence irrelevant.

3 See Alston (1993) for a detailed examination of epistemic circularity in the epistemic justification of perceptual beliefs. Epistemic circularity is not a major concern for those who abandon the internalist constraints on epistemic resources, but it remains a serious problem for epistemologists with an internalist conviction. See, for example, McGrew and McGrew (2007, Ch. 4).

4 We may write down everything we observed immediately to avoid reliance on memory, but then we need the reliable memory of linguistic rules when we write and read these notes.

5 Some people (e.g. Coady 1992) make the same point on testimonial evidence, viz. any (sufficiently strong) body of evidence in support of the reliability of testimony will include some evidence that comes from testimony, so that the reasoning in support of the reliability of testimony is epistemically circular. As noted in note 2, I do not discuss testimonial evidence because the notion of the ideal evaluator with an unlimited access to the epistemic resources of the community makes testimonial evidence irrelevant. See Shogenji (2006) for a response to the charge of epistemic circularity about the reliability of testimonial evidence.

6 Shogenji (2002) proposes two ways of confirming evidential independence without epistemic circularity with regard to *sense perceptual evidence*, but the two methods tacitly assume the reliability of memory.

7 That is the main focus in the recent literature on coherence, such as Bovens and Hartmann (2002) and Olsson (2005).

8 See Huemer (1997) and Olsson (2002) for formal proofs (they use slightly different conditions) against justification by coherence from scratch. While agreeing with their conclusion, Shogenji (2005) argues that justification by *recurrent coherence* from scratch is possible. The argument overcomes some obstacles of justification by coherence from scratch, but it is not effective against empirically equivalent alternative hypotheses. This chapter obtains the same result (justification up to empirical equivalence) with much simpler reasoning.

9 See Alston (1989, pp. 326–329, 1993, pp. 15–17) for the distinction between logical circularity and epistemic circularity.

10 A notable exception is Brown (1993, 1994).

11 I do not add the condition that $P(p_1 \wedge \ldots \wedge p_n) > 0$ to exclude the other degenerate case. It is not necessary because I am assuming, as usual, that $P(c \mid p_1 \wedge \ldots \wedge p_n)$ is undefined unless $P(p_1 \wedge \ldots \wedge p_n) > 0$. It follows from this assumption and the condition $P(c \mid p_1 \wedge \ldots \wedge p_n) = 1$ that $P(p_1 \wedge \ldots \wedge p_n) > 0$.

12 The term "likelihood" is used here in the technical sense that is standard in the literature, viz. the conditional probability $P(e \mid h \wedge b)$ of e given h (modulo b) is called "the likelihood of h".

13 The right to left direction of the last biconditional holds because $P(e \mid b)$ is the weighted average, $P(e \mid h \wedge b) \times P(h \mid b) + P(e \mid \neg h \wedge b) \times P(\neg h \mid b)$, of $P(e \mid h \wedge b)$ and $P(e \mid \neg h \wedge b)$.

14 I make this assumption about b to debunk the myth of epistemic circularity in a simple setting first. I will address the suspicion in the next section that epistemic circularity may be swept under the rug of the background information.

15 To make the case of testimony meaningful, the notion of the ideal evaluator is set aside here. The ideal evaluator needs no testimonial evidence because she has an unlimited access to the best epistemic recourses available to the community.

16 See, for example, Douven (2005, 2008).

17 So, empirical equivalence in the deductive sense does not imply empirical equivalence in the Bayesian sense. The implication in the opposite direction holds since we are assuming that the probability distribution is regular, so that $P(e \mid h \wedge b) = 1$ only if $\{h, b\} \vdash e$, and $P(e \mid h \wedge b) = 0$ only if $\{h, b\} \vdash \neg e$.

18 I am following the standard Bayesian format of epistemic evaluation here that we have good reason for accepting a proposition just in case its probability is sufficiently high. I will question this view in the next chapter. Also, the example mentions the background information b for the sake of generality, but we can make it empty in the Cartesian context.

19 As Peijnenburg and Atkinson (2013) put it, "[justification] gradually emerges from the chain [of reasons] as a whole" (p. 546).

20 And more generally with regard to any representational states, including beliefs and desires.

21 See Gordon (1995, 1996, 2007) and Byrne (2005) for similar views. Their proposals are general accounts of self-knowledge, but I confine my discussion to the reflective belief about the content of one's sensory state. Byrne (2012) applies his proposal to the self-knowledge of visual experience.

22 Byrne (2005) endorses the epistemic rule BEL "If *p*, believe that you believe that *p*" (p. 95), which allows us to form a reflective belief. The problem with this epistemic rule, for our present purpose, is that we must ascertain the truth of the antecedent *p* before we form the reflective belief.

23 There is a question of how the subject recognizes the modality of the sensory state. I cannot offer a full account here, but the most straightforward answer is the phenomenal content, i.e. the subject can distinguish visual experience from experiences of other sensory modalities by the phenomenal content of her experience.

24 Though the framework is devoid of the representational structure, it is not devoid of a certain structure. For example, the framework tacitly assumes the existence of an epistemic subject who is acquainted with the phenomenal qualities of her sensory states.

25 Evans (1982) stresses this point about mental states in general, while Harman (1990) advances it more specifically on sensory states.

4 Dual Components of Epistemic Evaluation

4.1 Bayesian Formats of Epistemic Evaluation

The previous chapter addressed the problem of epistemic circularity and the problem of epistemic regress, but solving these problems is not much help when it comes to Cartesian skepticism. This is because the core issue of Cartesian skepticism is not epistemic circularity or epistemic regress, but empirically equivalent alternative hypotheses. The natural world hypothesis h_{NW} and the BIV hypothesis h_{BIV} are empirically equivalent to each other in the sense that for any possible reflective sensory evidence e,[1] their likelihoods $P(e \mid h_{\mathrm{NW}})$ and $P(e \mid h_{\mathrm{BIV}})$ are the same. As a result, the ratio of their probabilities remains the same no matter what sensory experience e we have:

$$\frac{P(h_{\mathrm{NW}} \mid e)}{P(h_{\mathrm{BIV}} \mid e)} = \frac{P(e \mid h_{\mathrm{NW}})P(h_{\mathrm{NW}}) \,/\, P(e)}{P(e \mid h_{\mathrm{BIV}})P(h_{\mathrm{BIV}}) \,/\, P(e)} = \frac{P(h_{\mathrm{NW}})}{P(h_{\mathrm{BIV}})}$$

It appears as though the dispute over Cartesian skepticism comes down to the comparison of the a priori probabilities, i.e. what probabilities we assign to the two hypotheses h_{NW} and h_{BIV} prior to obtaining any sensory experience. If so, it would be helpful to have some principles that guide the assignment of a priori probabilities. This is a difficult task, and some Bayesians think that a priori probabilities are simply subjective.[2] If that is the case, our choice between the natural world hypothesis and the BIV hypothesis may be a matter of subjective judgment. That might be disappointing to some epistemologists, but perhaps not a disaster because most people's subjective judgment is in favor of the natural word hypothesis.

This diagnosis appears sensible on surface, but it actually misses the point of Cartesian skepticism. The Cartesian skeptic is not trying to show that the BIV hypothesis is preferable to the natural world hypothesis. The skeptic is only urging us to withhold judgment. It is not even necessary for the skeptic to show that the two hypotheses are equally preferable. The skeptic only asserts that accepting only the disjunction of the two hypotheses is preferable to accepting the natural world hypothesis in exclusion of the BIV hypothesis. Indeed, unless the BIV hypothesis is refuted with absolute certainty, the probability $P(h_{NW} \vee h_{BIV} \mid e)$ is higher than the probability $P(h_{NW} \mid e)$ given any collection of sensory experience *e*. Of course, it is possible that both $P(h_{NW} \vee h_{BIV} \mid e)$ and $P(h_{NW} \mid e)$ are sufficiently high for acceptance in the sense that $P(h_{NW} \vee h_{BIV} \mid e) \geq t$ and $P(h_{NW} \mid e) \geq t$ for some threshold of sufficiency *t*. However, defending the claim $P(h_{NW} \mid e) \geq t$ is more difficult than defending the claim $P(h_{BIV} \mid e) < P(h_{NW} \mid e)$. More importantly, there is something odd about the idea that accepting the disjunction of the competing hypotheses is always preferable to accepting one of them (unless the other hypothesis is refuted with absolute certainty). I am going to argue against the standard Bayesian format of epistemic evaluation that has this consequence.

It is necessary to spell out, first, the basic features of the Bayesian evaluation of hypotheses. As mentioned earlier, there are two types of epistemic evaluation. One is qualitative and the other is quantitative. Qualitative epistemic evaluation determines whether or not we can accept a hypothesis in light of evidence, e.g. whether or not we can accept the hypothesis that there will be rain tomorrow in light of the weather forecast.[3] The evaluation is meant to help us increase true beliefs and avoid false beliefs. Quantitative evaluation, meanwhile, assigns a probability to the hypothesis, e.g. I may assign the probability 0.8 to the hypothesis that there will be rain tomorrow in light of the weather forecast. The goal in quantitative evaluation is not increasing true beliefs and avoiding false beliefs because regardless of the actual weather tomorrow, I do not have a true belief if I only believe the hypothesis with the degree 0.8. The full belief that the probability of the hypothesis being true is 0.8 is also false—only the belief that the probability is 0 or the belief that the probability is 1 can be true. Once the actual weather is revealed, we evaluate the probability assignment by its closeness to the truth, e.g. if there is rain tomorrow, the probability 0.8 is closer to the truth than is the probability 0.7.

This point about quantitative evaluation holds even when the probability is interpreted as the frequency ratio in the long run, e.g. the probability 0.8 may mean that given the type of condition that holds today, the ratio of rain next day is 8 out of 10 in the long run. Note that the estimate is almost always wrong in the long run. For example, the estimate of 0.8 is wrong if the true ratio is 80,001 out of 100,000 instead of 80,000 out of 100,000. It is foolhardy to believe in a precise ratio such as 0.8 if the goal is to increase true beliefs and avoid false beliefs. So, even under the frequency interpretation, the goal of quantitative evaluation is not increasing true beliefs and avoiding false beliefs, but to assign a probability that is close to the true value.

It is commonly understood that Bayesianism evaluates a hypothesis in a quantitative way by assigning a probability to it. Though this is true, there are two important caveats. First, a Bayesian format of epistemic evaluation does not actually measure the epistemic standing of the hypothesis directly from the body of evidence. Take the standard Bayesian format that measures the degree of epistemic justification by the conditional probability $P(h \mid e \wedge b)$. Bayes's Theorem tells us how to calculate $P(h \mid e \wedge b)$ from the three probabilities $P(e \mid h \wedge b)$, $P(h \mid b)$ and $P(e \mid b)$.[4] However, Bayes's Theorem—or any law of probability theory—does not tell us how to determine these three probabilities themselves. The determination of the probabilities from the body of evidence requires a format of a different kind.[5] A Bayesian format of epistemic evaluation only relates the degree of epistemic justification to some probabilities that are already determined.

The second caveat is that a quantitative evaluation is often a step in the process of obtaining a qualitative evaluation. A common practice among Bayesians is to introduce a threshold of sufficiency t for the acceptance of a hypothesis. The resulting format is that there is justification for accepting the hypothesis h given the evidence e if and only if $P(h \mid e \wedge b) \geq t$. I will call this format "Lockean" as is customary in the literature.[6] The Lockean format is Bayesian, but it is a format of qualitative evaluation. Other Bayesian formats of quantitative evaluation can be turned into qualitative evaluation in the same way by the introduction of some threshold of sufficiency.

A remark is in order here on one feature of epistemic evaluation in meliorative *social* epistemology, viz. there is a distinction between the evaluator and the epistemic subject whose beliefs are evaluated. Epistemic evaluation in meliorative social epistemology is constrained by resources available to the community, and I will continue the idealization introduced earlier that the evaluator has an unlimited access to the best epistemic recourses—both empirical and conceptual—available to the community. Meanwhile, it is the beliefs of the individual epistemic subject with limited epistemic resources that are evaluated. So, qualitative epistemic evaluation is meant to help the subject to increase true beliefs and avoid false beliefs, where the amount of increase in true beliefs is relative to the body of beliefs already held by the subject. Of course, the body of beliefs varies among the members of the community, but we can measure the amount of increase in true beliefs relative to the body of statements the evaluator has already endorsed for the community to accept.[7] In other words, we consider the ideal epistemic subject who accepts exactly the statements endorsed by the evaluator, and measure the amount of increase in true beliefs relative to the body of beliefs held by the ideal subject. Hereafter, "b" refers to the body of beliefs in this sense, while the evidence e includes all information available to the evaluator beyond the body of beliefs b.

4.2 The Backdoor Endorsement Problem of the Lockean Format

This section argues against the Lockean format of epistemic evaluation, according to which there is justification for accepting the hypothesis h just in case $P(h \mid e \wedge b) \geq t$, where e is the body of evidence available to the evaluator, b is the body of beliefs held by the subject and t is a threshold of sufficiency. The Lockean format is grounded in the standard Bayesian format of epistemic evaluation, which measures the degree of justification by $P(h \mid e \wedge b)$. I already voiced my concern about the measure in the previous section, viz. barring refutation with absolute certainty, the format favors remaining non-committal between any pair of competing hypotheses h_1 and h_2 over accepting one of them in exclusion of the other. This is because $P(h_1 \mid e \wedge b) < P(h_1 \vee h_2 \mid e \wedge b)$ and $P(h_2 \mid e \wedge b) < P(h_1 \vee h_2 \mid e \wedge b)$ unless h_1 or h_2 is refuted with absolute certainty. This section turns this problem into a formal argument against the Lockean format of epistemic evaluation.

It is well known that justification by the Lockean format is not closed under logical entailment from multiple premises.[8] In other words, even if there is justification for accepting each of the multiple premises $p_1, \ldots, p_n$, and $\{p_1, \ldots, p_n\}$ logically entails the hypothesis h, there may still be no justification for accepting h. To put this formally, it does not follow from $P(p_i \mid e \wedge b) \geq t$ for $i = 1, \ldots, n$ and $\{p_1, \ldots, p_n\} \vdash h$, that $P(h \mid e \wedge b) \geq t$. To see this in a simple example, let h be the conjunction of the two premises p_1 and p_2, and thus $\{p_1, p_2\} \vdash h$. Suppose also that $P(p_1 \mid e \wedge b) \geq t$ and $P(p_2 \mid e \wedge b) \geq t$ for some threshold t. It is still possible that $P(h \mid e \wedge b) < t$ in violation of closure because $P(h \mid e \wedge b) = P(p_1 \wedge p_2 \mid e \wedge b) < P(p_1 \mid e \wedge b)$ and $P(h \mid e \wedge b) = P(p_1 \wedge p_2 \mid e \wedge b) < P(p_2 \mid e \wedge b)$ unless p_1 or p_2 is refuted with absolute certainty.

The failure of multiple-premise closure gives rise to the problem of the backdoor endorsement, viz. even if the evidence for h falls short of justifying an endorsement of h, the evaluator can endorse it through the backdoor by endorsing the premises $p_1, \ldots, p_n$, for each of which the evidence is sufficiently strong, and letting the subject derive h from $\{p_1, \ldots, p_n\}$. This problem is commonly known in the form of the lottery paradox.[9] Suppose each lottery ticket has the number 1, 2, ... or n on it for some large number n, and exactly one number is the winner. Each premise p_i states that number i is *not* the winning number, while the conclusion h is that none of the tickets has the winning number. The probability of each premise is close to one because n is large. So, the evaluator can endorse the conclusion h—which is false in this setting—through the backdoor by endorsing each of the premises $p_1, \ldots, p_n$ and letting the subject derive h from $\{p_1, \ldots, p_n\}$.

The lottery paradox helps us see the problem of backdoor endorsement, but it is in need of refinement because as it stands, the lottery paradox has a solution that is consistent with the Lockean format. The key to the solution is changes in the body of beliefs due to the series of endorsements. Let b be the initial body of beliefs. As the evaluator endorses p_1, p_2, ... to the epistemic subject, the initial body b expands to $b \wedge p_1$, $b \wedge p_1 \wedge p_2$, ..., so that by the time the evaluator considers the last premise p_n, the body of beliefs is $b \wedge p_1 \wedge \ldots \wedge p_{n-1}$. Given this expanded body of beliefs, there is no justification for endorsing p_n in the Lockean format because $P(p_n \mid b \wedge p_1 \wedge \ldots \wedge p_{n-1}) = 0$.[10] In fact, the point comes earlier in the series of endorsements where $P(p_{i+1} \mid b \wedge p_1 \wedge .... \wedge p_i) < t$ and thus there is no justification for endorsing the next premise p_{i+1}. In other words, it is not possible to endorse all of them because at some point in the series of endorsements the expanded body of beliefs blocks the endorsement of the next premise. There is therefore no backdoor endorsement of h because the subject needs all of the premises $\{p_1, ..., p_n\}$ to derive the conclusion h.[11]

The solution keeps the Lockean format intact, but the strategy does not carry over to all cases of backdoor endorsement. One feature of the lottery paradox is crucial to the solution, viz. as the body of beliefs expands from b to $b \wedge p_1$, to $b \wedge p_1 \wedge p_2$, and so on, the conditional probability of the next premise $P(p_{i+1} \mid b \wedge p_1 \wedge .... \wedge p_i)$ steadily declines, and eventually reaches zero at $P(p_n \mid b \wedge p_1 \wedge .... \wedge p_{n-1}) = 0$. It is this feature that makes it impossible for the evaluator to endorse all the premises $p_1, ..., p_n$ in the lottery case. The solution does not carry over to cases without this feature.

Consider the following variant of the lottery paradox. As before, each lottery ticket has the number 1, 2, ... or n on it for some large number n, and exactly one number is the winner. This time, however, each premise p_i states that the i-th *draw* does not have the winning number on it, and for any ticket drawn, the pool is replenished with a new ticket with the same number, so that the probability of the premise is $P(p_i \mid b) = (n-1)/n > t$ for any i. The premises in the case are $p_1, ..., p_m$, where m need not be n, while the conclusion h is that none of the first m draws has the winning number on it. The probability of the conclusion is therefore $P(h \mid b) = P(p_1 \wedge ... \wedge p_m \mid b) = [(n-1)/n]^m$. We assume that the evaluator cannot endorse h to the subject because m is sufficiently large to make $P(h \mid b) = [(n-1)/n]^m < t$. The situation is similar to the standard version of the lottery paradox in that the evaluator can endorse each of the premises $p_1, ..., p_m$ on its own, but not the conclusion they logically entail. The difference, however, is that as the body of beliefs expands from b to $b \wedge p_1$, to $b \wedge p_1 \wedge p_2$, and so on, the conditional probability of the next premise does not become smaller. It remains the same at $P(p_{i+1} \mid b \wedge p_1 \wedge .... \wedge p_i) = (n-1)/n \geq t$ for any i. This means that the evaluator can endorse all of the premises $p_1, ..., p_m$ in sequence.[12] The problem of backdoor endorsement is back again, i.e. the evaluator can endorse all of the premises $p_1, ..., p_m$ and let the subject derive h from $\{p_1, ..., p_m\}$.

We can generalize the point. First, epistemic evaluation in the lottery paradox involves no evidence e beyond the initial body of beliefs b, but that has no bearing on the problem of backdoor endorsement. Where there is some body of evidence e, we can simply replace b by the conjunction $b \wedge e$. Second, we can establish the general claim that for any hypothesis h whose probability falls short of the threshold t of sufficiency, there is a set of premises such that on condition of $e \wedge b$, (1) each of them has the probability t or higher, (2) they are probabilistically independent of each other, and (3) they jointly entail h. The three conditions allow the evaluator to endorse h through the backdoor in the way described. Here is an instruction to find such a set of premises for any hypothesis h. Let $p_1, \ldots, p_n$ be the disjunctions $h \vee d_1, \ldots, h \vee d_n$ such that (4) $\{d_1, \ldots, d_n\}$ is inconsistent given $\neg h$ in that $P(d_1 \wedge \ldots \wedge d_n \mid \neg h \wedge e \wedge b) = 0$, (2) $p_1, \ldots, p_n$ are probabilistically independent of each other so that $P(p_2 \mid p_1 \wedge e \wedge b) = P(p_2 \mid e \wedge b), \ldots, P(p_{i+1} \mid p_1 \wedge \ldots \wedge p_i \wedge e \wedge b) = P(p_{i+1} \mid e \wedge b), \ldots, P(p_n \mid p_1 \wedge \ldots \wedge p_{n-1} \wedge e \wedge b) = P(p_n \mid e \wedge b)$, and (5) $p_1, \ldots, p_n$ are equally probable in that $P(p_1 \mid e \wedge b) = \ldots = P(p_n \mid e \wedge b)$.[13] It follows from (4) that h is logically equivalent to the conjunction $p_1 \wedge p_2 \wedge \ldots \wedge p_n$, and thus (3) $\{p_1, \ldots, p_n\}$ logically entails h. It also follows from (2) and (4) that $P(h \mid e \wedge b) = P(p_1 \wedge \ldots \wedge p_n \mid e \wedge b) = P(p_1 \mid e \wedge b) \times \ldots \times P(p_n \mid e \wedge b)$. It follows further in combination with (5) that $P(p_1 \mid e \wedge b) = \ldots = P(p_n \mid e \wedge b) = P(h \mid e \wedge b)^{1/n}$. So, as n increases, $P(p_1 \mid e \wedge b) = \ldots = P(p_n \mid e \wedge b) = P(h \mid e \wedge b)^{1/n}$ also increases and approaches one for any h such that $0 < P(h \mid e \wedge b) < 1$. We can therefore make the probability of each premise sufficiently high in that (1) $P(p_1 \mid e \wedge b) = \ldots = P(p_n \mid e \wedge b) \geq t$ by choosing sufficiently large n. The upshot of the construction is that for any hypothesis h whose probability falls short of the threshold t of sufficiency, the evaluator can endorse h through the backdoor. The Lockean format of epistemic evaluation is in serious trouble.

One possible response is to defend the Lockean format by some restriction that blocks the backdoor endorsement, but there is no sensible way to do so. If there is justification for endorsing each of $p_1, \ldots, p_m$ and their evaluations are not affected by the expansion of the body of beliefs, then there is justification for endorsing all of them in sequence. If we still wish to prevent the subject from accepting h, we must block the logical inference from $\{p_1, \ldots, p_m\}$ to h. There is no good reason for such a restriction. There is also the draconian response of giving up the notion of qualitative cognitive attitudes altogether in favor of the quantitative attitudes. In other words, we never accept a hypothesis, reject it or withhold judgment. Instead we believe any statement to a degree, e.g. believing the natural world hypothesis to a certain degree and believing the BIV hypothesis to a certain degree. If we pursue this option, increasing true beliefs and avoiding false beliefs is no longer a goal of our epistemic evaluation. We will only be pursuing a quantitative goal, e.g. to make our degree of belief as close as possible to the true value. I will consider this option if all plausible ways of saving qualitative cognitive attitudes fail, but I do not think we are there yet.[14]

One of the less draconian options is to retain *qualitative cognitive attitudes* but abandon the *classification* of hypotheses. We can do so by understanding the standard Bayesian measure $P(h \mid e \wedge b)$ as the degree of epistemic justification for accepting (fully believing) the hypothesis h, instead of the degree of belief, but without the threshold of sufficiency t that classifies hypotheses into those we can accept and those we cannot accept. This is not a radical departure from the Lockean format because we can still compare the degrees of epistemic justification among competing hypotheses, such as the natural world hypothesis and the BIV hypothesis.

This option is still suspect for the very reason that it retains the standard Bayesian measure $P(h \mid e \wedge b)$ of epistemic justification. As mentioned already, it has the questionable consequence of favoring the attitude of never choosing between any pair of competing hypotheses h_1 and h_2 unless one of them is refuted with absolute certainty because $P(h_1 \mid e \wedge b) < P(h_1 \vee h_2 \mid e \wedge b)$ and $P(h_2 \mid e \wedge b) < P(h_1 \vee h_2 \mid e \wedge b)$. In a way, dropping the threshold of sufficiency t makes the situation worse. It is no longer possible to maintain that we can accept a disjunct, say h_1, in addition to the disjunction $h_1 \vee h_2$ because $P(h_1 \mid e \wedge b) \geq t$ in addition to $P(h_1 \vee h_2 \mid e \wedge b) \geq t$. In the absence of a threshold, there is no good account of why we can also accept the disjunct whose justification is weaker than that of the disjunction. It is sensible to explore other options, and I will take up the option of abandoning the standard Bayesian measure and replacing it with a new measure.

4.3 The Risk and Gain in Epistemic Evaluation

This section presents an outline of a new Bayesian format of epistemic evaluation, which measures the "epistemic worthiness" of a hypothesis given the body of evidence e and the body of beliefs b.[15] The specific measure $J(h, e \mid b)$ of epistemic worthiness will be proposed and defended in Section 4.

It was pointed out in the previous section that the standard Bayesian measure $P(h \mid e \wedge b)$ has a questionable consequence, and that the Lockean format of epistemic evaluation derived from it is in serious trouble. My diagnosis of the problem is as follows: $P(h \mid e \wedge b)$ is a good measure of how safe it is for the subject to accept h, but the epistemic worthiness of acceptance does not solely depend on the degree of safety. It also depends on the potential gain in truth by the acceptance of h. For example, accepting the conjunction $p_1 \wedge p_2$ is riskier than accepting only p_1 or only p_2 unless one of the conjuncts is redundant.[16] It is therefore sensible to accept only p_1 or only p_2 if safety is our only concern. However, if we accept the conjunction $p_1 \wedge p_2$ instead of only p_1 or only p_2, the potential gain in truth is greater. Similarly, in the evaluation of the competing hypotheses h_1 and h_2, accepting one of them in exclusion of the other is riskier than accepting their disjunction $h_1 \vee h_2$, but if we accept one of them, the potential gain in truth is greater. In evaluating the epistemic worthiness of a hypothesis, we need to balance the risk and the potential gain because it is worth taking a greater risk when the potential gain is greater. The standard Bayesian measure $P(h \mid e \wedge b)$ only takes into account the safety of accepting h, and fails to balance it with the potential gain in truth.

It may appear the Lockean format addresses this issue by setting the threshold of sufficiency t at less than one. If our only concern is safety, then we should return to the Cartesian requirement of absolute certainty by setting the threshold of sufficiency t at one. The Lockean format sacrifices some safety for potential gain by allowing us to accept a hypothesis that is less than certain. The problem, however, is that the Lockean format does not take into account the amount of potential gain in truth. Consider a large volume on the history of China packed with specific and detailed information. Even if the author is highly competent and very careful, the probability that at least one of the factual statements is false is substantial, but a large amount of potential gain in truth can make their conjunction worthy of acceptance. This is very different from accepting a broad and unspecific statement that tells us little, e.g. there will be a recession in China sometime in the next decade, even if the degree of safety is the same. It is worth taking a greater risk if the amount of potential gain in truth is greater, but the Lockean format ignores the difference and applies the same threshold t regardless of the amount of potential gain in truth. My goal in this chapter is to formulate a format of epistemic evaluation that balances the amount of risk and the amount of potential gain.

Before taking on the task of balancing them, we must settle on the measure of safety and the measure of potential gain. Of these two, the measure of safety is straightforward, viz. the risk of accepting falsity is a decreasing function of $P(h \mid e \wedge b)$. Since we prefer greater safety—a lower risk of accepting falsity—the epistemic worthiness of the hypothesis h increases as $P(h \mid e \wedge b)$ becomes higher, other things being equal. The probability $P(h \mid e \wedge b)$ itself is the most natural measure of safety, though any increasing function of $P(h \mid e \wedge b)$ will serve the purpose. Meanwhile, it is not as straightforward to measure the potential gain in truth. It is noted in the examples given earlier that the potential gain in truth is greater if we accept the conjunction $p_1 \wedge p_2$ instead of only p_1 or only p_2, and that the potential gain in truth is greater if we accept h_1 on its own or h_2 on its own, instead of the disjunction $h_1 \vee h_2$. However, we cannot simply count the number of conjuncts and disjuncts to measure the potential gain in truth. The same content can be articulated in different ways with different numbers of conjuncts and disjuncts in different languages.

A more sensible approach is to measure the potential gain in truth by the amount of information the statement carries. Seeking a greater amount of information is a familiar theme in philosophy of science. Here is one representative statement:

> *Science does not aim, primarily, at high probabilities. It aims at a high informative content, well backed by experience. But a hypothesis may be very probable simply because it tells us nothing, or very little.* A high degree of probability is therefore not an indication of "goodness"—it may be merely a symptom of low informative content.
>
> (Popper 1954, p. 146, original italics)

In Popper's case the emphasis on informativeness is accompanied by pessimism about the truth of scientific theories (universal statements), but we can accept informativeness as a worthy aim without abandoning the aim of truth.[17] If we extend the emphasis on informativeness beyond the evaluation of a scientific hypothesis, then whether any statement is worthy of acceptance depends not only on the risk that it is false, but also on the amount of information the statement carries.[18]

The next question is then how we measure the amount of information. The most familiar measure is a logarithmic function of the probability (Shannon and Weaver 1949). If we adopt 2 as the base of logarithm, the statement *h* carries $I(h) = -\log_2 P(h)$ bits of information. A bit is a convenient unit since the storage space for binary-coded information (e.g. written in 0s and 1s) is proportional to the amount of stored information in bits. For example, if the probability of the statement *h* is 1/8, then *h* carries 3 bits of information since $I(h) = -\log_2 1/8 = -\log_2 2^{-3} = 3$. So, we need three units of storage space (for three binary digits). Each time the probability is reduced by half, we need one additional unit of storage space since the statement carries one more bit of information. For example, $I(h) = -\log_2 1/16 = 4$ if $P(h) = 1/16$; and $I(h) = -\log_2 1/32 = 5$ if $P(h) = 1/32$, etc. We can replace the log base 2 with some other number greater than one, such as 10 ("dit") or Euler's number *e* ("nat").[19] The difference among them is conceptually insignificant, but there are more interesting alternatives. It has been suggested (Popper 1963, p. 392; Bar-Hillel and Carnap 1953, among others) that the amount of information is the subtraction of the probability from unity, i.e. $1 - P(h)$. Another possibility mentioned (Levi 2004, Section 4.1) is the reciprocal of the probability, i.e. $1/P(h)$.[20]

All these proposals have one feature in common, viz. they are all inverse functions of the probability $P(h)$. We can see the reason for this in two steps. First, the degree of specificity is directly (positively) related to the amount of information. The more specifically the statement describes the situation, the larger amount of information it carries. Second, the degree of specificity is inversely related to the probability of the statement. The more specifically the statement describes the situation, the lower the probability is that it is true. By combining these two steps, we see that the amount of information the statement carries is inversely related to its probability. We need not commit ourselves yet to any particular measure of information in exclusion of others. For now, I only state that the amount of information is inversely related to the probability of the statement.

A more pressing issue is which of *h*'s probabilities is appropriate for measuring the amount of information *h* carries. There are three candidates: $P(h)$, $P(h \mid b)$ and $P(h \mid e \wedge b)$. Some propose that the *a priori* probability of the hypothesis determines the amount of information (Bar-Hillel and Carnap 1953). On this view the amount of information *h* carries is inversely related to the probability $P(h)$. Others propose that the *epistemic* probability of the hypothesis determines the amount of information (Hempel 1960). On this view the amount of information *h* carries is inversely related to the probability $P(h \mid e \wedge b)$.[21] Both proposals have merits and are useful for different purposes, but neither of them is appropriate in the context of meliorative social epistemology, where the evaluator judges whether or not to endorse a hypothesis to the epistemic subject. Since the endorsement is meant to help the epistemic subject to increase true beliefs and avoid false beliefs, the amount of increase in true beliefs is relative to the body of beliefs *b* held by the epistemic subject. For example, if *h* is already in the epistemic subject's body of beliefs *b*, then adding *h* to it does not increase true beliefs. If, on the other hand, *h* is unexpected given the epistemic subject's body of beliefs *b*, then adding *h* to *b* increases true beliefs greatly. So, if we use the amount of information that *h* carries in measuring the increase in true beliefs, then the amount of information that *h* carries should be inversely related to the probability $P(h \mid b)$ instead of the a priori probability $P(h)$ or the epistemic probability $P(h \mid e \wedge b)$.

4.4 The Dual-Component Format of Epistemic Evaluation

The next step in the formulation of a new Bayesian measure of epistemic evaluation is to find the right balance between safety and informativeness. I have argued so far that the epistemic worthiness of the hypothesis *h* increases as the safety of *h* increases (the risk of *h*'s falsity decreases) and as the amount of information *h* carries increases. To put this formally, the measure $W(h, e \mid b)$ of the epistemic worthiness of *h* is an increasing function of $P(h \mid e \wedge b)$ and a decreasing function of $P(h \mid b)$, where *e* is the body of evidence available to the evaluator, and *b* is the body of the subject's beliefs.

There are, however, numerous functions that meet these requirements, and many of them are not ordinally equivalent to each other.[22] Take $D(h, e \mid b)$ and $R(h, e \mid b)$ below:

$$D(h, e \mid b) = P(h \mid e \wedge b) - P(h \mid b)$$

$$R(h, e \mid b) = \frac{P(h \mid e \wedge b)}{P(h \mid b)}$$

Both functions meet the two requirements, but they are not ordinally equivalent to each other, i.e. there is some pair of hypotheses h_1 and h_2 such that their order is different by $D(h, e \mid b)$ and by $R(h, e \mid b)$. For example, suppose $P(h_1 \mid b) = 0.1$ and $P(h_1 \mid e \wedge b) = 0.4$, while $P(h_2 \mid b) = 0.2$ and $P(h_2 \mid e \wedge b) = 0.6$. It is easy to see that $D(h_1, e \mid b) < D(h_2, e \mid b)$ while $R(h_1, e \mid b) > R(h_2, e \mid b)$, as follows:

$$D(h_1, e \mid b) = P(h_1 \mid e \wedge b) - P(h_1 \mid b) = 0.4 - 0.1 = 0.3$$

$$D(h_2, e \mid b) = P(h_2 \mid e \wedge b) - P(h_2 \mid b) = 0.6 - 0.2 = 0.4$$

$$R(h_1, e \mid b) = \frac{P(h_1 \mid e \wedge b)}{P(h_1 \mid b)} = \frac{0.4}{0.1} = 4$$

$$R(h_2, e \mid b) = \frac{P(h_2 \mid e \wedge b)}{P(h_2 \mid b)} = \frac{0.6}{0.2} = 3$$

The task of this section is to find a measure epistemic worthiness $W(h, e \mid b)$ that balances the two determinants, $P(h \mid e \wedge b)$ and $P(h \mid b)$, in the right way.

Fortunately, we already have a significant formal constraint on $W(h, e \mid b)$ from Section 2.[23] Recall the variant of the lottery case where the premises $p_1, \ldots, p_m$ are irrelevant to each other (probabilistically independent of each other) so that the expansion of the initial body of beliefs from b to $b \wedge p_1$, to $b \wedge p_1 \wedge p_2$, and so on does not affect the epistemic evaluation of any premise that remains. The Lockean format of epistemic evaluation runs into trouble when $P(p_1 \mid b) \geq t, \ldots, P(p_m \mid b) \geq t$ but $P(p_1 \wedge \ldots \wedge p_m \mid b) < t$ because even though there is no justification for endorsing $p_1 \wedge \ldots \wedge p_m$ by the Lockean format, the evaluator can still endorse it through the backdoor, i.e. by endorsing all of the premises $p_1, \ldots, p_m$ in sequence and letting the subject derive $p_1 \wedge \ldots \wedge p_m$ from them. An adequate format of epistemic evaluation should not allow backdoor endorsement of this kind. So, if $p_1, \ldots, p_m$ are irrelevant to each other and there is no justification for endorsing their conjunction, then it should not be the case that there is justification for endorsing each of them. Equivalently, if $p_1, \ldots, p_m$ are irrelevant to each other and there is justification for endorsing each of them, then there should be justification for endorsing their conjunction.

When I argued against the Lockean format in Section 2, the argument is not affected by the inclusion of some body of evidence e. What counts for the argument is the probabilistic independence of $p_1, \ldots, p_m$ on condition of $e \wedge b$, where e can be empty. This point no longer applies when we try to balance the degree of safety, which is a function of $P(h \mid e \wedge b)$, and the amount of information, which is a function of $P(h \mid b)$. To ensure that $W(h_1, e \mid b) \geq t, \ldots, W(h_n, e \mid b) \geq t$ allow the endorsement of all of $h_1, \ldots, h_n$ in sequence, $h_1, \ldots, h_n$ must be irrelevant to each other both in terms of safety and information. To state this formally, $h_1, \ldots, h_n$ must be probabilistically independent both on condition of b and on condition of $e \wedge b$, so that $P(h_i \mid b \wedge h_1 \wedge \ldots \wedge h_{i-1}) = P(h_i \mid b)$ and $P(h_i \mid e \wedge b \wedge h_1 \wedge \ldots \wedge h_{i-1}) = P(h_i \mid e \wedge b)$ for any $i = 2, \ldots, n$. It follows from this condition that $W(h_i, e \mid b \wedge h_1 \wedge \ldots \wedge h_{i-1}) = W(h_i, e \mid b)$ for any $i = 2, \ldots, n$ because $W(h, e \mid b)$ is determined by $P(h \mid e \wedge b)$ and $P(h \mid b)$. So, under the strengthened condition of irrelevance, $W(h_{i+1}, e \mid b) \geq t$ for $i = 1, \ldots, n$ allows the evaluator to endorse the conjunction $h_1 \wedge \ldots \wedge h_n$ through the backdoor by endorsing all of $h_1, \ldots, h_n$ in sequence and letting the subject derive $h_1 \wedge \ldots \wedge h_n$. Here is then a formal constraint on an adequate measure of epistemic worthiness:

> **Backdoor Constraint:** If the statements $h_1, \ldots, h_n$ are probabilistically independent of each other, both on condition of b and on condition of $b \wedge e$, and if $W(h_1, e \mid b) \geq t, \ldots, W(h_n, e \mid b) \geq t$, then $W(h_1 \wedge \ldots \wedge h_n, e \mid b) \geq t$.

The backdoor constraint is applicable to special cases where $P(h \mid e \wedge b) = P(h \mid b)$, i.e. even if there is no evidence e beyond the body of beliefs b as in the variant of the lottery case used against the Lockean format.

The Backdoor Constraint has a negative counterpart, viz. if the statements $h_1, \ldots, h_n$ are irrelevant to each other, and if there is no justification for endorsing any of them, then there is no justification for endorsing their conjunction $h_1 \wedge \ldots \wedge h_n$. Here is a formal statement:

> **Inverse Backdoor Constraint:** If the statements $h_1, \ldots, h_n$ are probabilistically independent of each other, both on condition of b and on condition of $b \wedge e$, and if $W(h_1, e \mid b) < t, \ldots, W(h_n, e \mid b) < t$, then $W(h_1 \wedge \ldots \wedge h_n, e \mid b) < t$.

Without this constraint, the evaluator can endorse all of $h_1, \ldots, h_n$ through the backdoor by endorsing $h_1 \wedge \ldots \wedge h_n$ and let the subject derive the conjuncts from it. It is convenient to have a name for the combination of the Backdoor Constraint and the Inverse Backdoor Constraint. I will call them "the General Conjunction Requirement" (GCR). So, an adequate measure $W(h, e \mid b)$ of epistemic worthiness must be (1) an increasing function of $P(h \mid e \wedge b)$, (2) a decreasing function of $P(h \mid b)$ and (3) a function that satisfies GCR.

It is helpful to make some consequences of GCR explicit. The most important among them is the Special Conjunction Requirement (SCR):

> **Special Conjunction Requirement:** If the statements $h_1, \ldots, h_n$ are probabilistically independent of each other, both on condition of b and on condition of $b \wedge e$, and if $W(h_1, e \mid b) = \ldots = W(h_n, e \mid b) = w$, then $W(h_1 \wedge \ldots \wedge h_n, e \mid b) = w$.

We can establish that GCR entails SCR by proving its contraposition. Suppose $W(h, e|b)$ fails to satisfy SCR, and thus for some statements $h_1, \ldots, h_n$ that are probabilistically independent of each other both on condition of b and on condition of $b \wedge e$, $W(h_1, e \mid b) = \ldots = W(h_n, e \mid b) = w$ but $W(h_1 \wedge \ldots \wedge h_n, e \mid b) = w + \alpha$ for some $\alpha \neq 0$. $W(h, e \mid b)$ then violates GCR as follows. If $\alpha > 0$, then the Inverse Backdoor Constraint is violated for the threshold $t = w + \alpha$ because $W(h_1, e \mid b) = \ldots = W(h_n, e \mid b) = w < t$ while $W(h_1 \wedge \ldots \wedge h_n, e \mid b) = w + \alpha \geq t$. If $\alpha < 0$, then the Backdoor Constraint is violated for the threshold $t = w$ because $W(h_1, e \mid b) = \ldots = W(h_n, e \mid b) = w \geq t$ while $W(h_1 \wedge \ldots \wedge h_n, e \mid b) = w + \alpha < t$. Since the denial of SCR entails the denial of GCR, GCR entails SCR.

SCR itself has two further consequences of significance: equi-neutrality and equi-maximality. To begin with the former, any continuous function $W(h, e \mid b)$ of $P(h \mid e \wedge b)$ and $P(h \mid b)$ that satisfies SCR is constant when $P(h \mid e \wedge b) = P(h \mid b)$ (see Appendix 1 for proof). This means that if the body of evidence e is neutral to the probability of the hypothesis h (on condition of the body of beliefs b) then the degree of epistemic worthiness is constant, regardless of $P(h \mid b)$. If we set the constant value at zero, then $W(h, e \mid b) >$ (or $=$, $<$) 0 if and only if $P(h \mid e \wedge b) >$ (or $=$, $<$) $P(h \mid b)$, respectively, because $W(h, e \mid b)$ is an increasing function of $P(h \mid e \wedge b)$ and a decreasing function of $P(h \mid b)$.[24] Some remarks are in order here since the equi-neutrality requirement may look suspect. Suppose h_1 is almost certainly true while h_2 is almost certainly false (given the body of beliefs b). If the body of evidence e is neutral to both h_1 and h_2, are they equally worthy of inclusion in the body of beliefs? It may look suspect initially but the answer is yes, and the reason is again the distinction between the degree of safety and the degree of epistemic worthiness. Adding h_1 to b is much safer than adding h_2 to b in the case, but it does not follow that h_1 is more worthy of inclusion in the body of beliefs than h_2 is because the greater safety of adding h_1 is offset by the smaller gain in truth we will make if h_1 turns out to be true. Since the degree of epistemic worthiness is to serve the dual goal of cognition, it is not unreasonable to assign the same degree of epistemic worthiness to h_1 and h_2.

The equi-neutrality requirement is helpful in our search for an adequate measure of epistemic worthiness because it points us to a family of measures that have been extensively studied in the literature, viz. measures of incremental confirmation.[25] It is generally agreed that the degree of incremental confirmation is equi-neutral, i.e. if the evidence does not affect the probability of the hypothesis, then there is no incremental confirmation, positive or negative. It is also generally agreed that the degree of incremental confirmation is an increasing function of $P(h \mid e \wedge b)$ and a decreasing function of $P(h \mid b)$. These are also requirements that a measure of epistemic worthiness should satisfy. So, we can examine various measures of incremental confirmation that are already available in the literature to find out whether some of them serve as measures of epistemic worthiness.

Of course, an argument used for or against a particular measure of incremental confirmation in the literature cannot be considered an argument for or against a particular measure of epistemic worthiness. Though they happen to share important features, the two measures are conceptually very different. The degree of incremental confirmation is the extent to which a new piece of evidence e makes the hypothesis h safer to accept, and the prior probability $P(h \mid b)$ is in the measure because $P(h \mid b)$ is the original degree of safety. Meanwhile, the prior probability $P(h \mid b)$ in the measure of epistemic worthiness is inversely related to the amount of information that h carries. It is not included to account for the (original) degree of safety. Since the two measures are conceptually very different, an adequate measure of epistemic worthiness may not have any additional features that are desirable for a measure of incremental confirmation.[26]

So, a measure of epistemic worthiness must be selected on its own merits based on the formal constraints, and it turns out we can narrow down the search further by another significant consequence of SCR, viz. an adequate measure of epistemic worthiness that satisfies SCR is not only equi-neutral but also equi-maximal. It does not simply mean that for any given $P(h \mid b)$, $W(h, e \mid b)$ should be the highest when $P(h \mid e \wedge b) = 1$. That is obvious because $W(h, e \mid b)$ is an increasing function of $P(h \mid e \wedge b)$. The equi-maximality requirement states further that this highest value should be constant, regardless of $P(h \mid b)$. This means that when the evidence e together with b makes the statement h certain, it is worthy of inclusion in our body of beliefs to the highest possible degree. So, there is justification for adding h to b no matter how high the threshold of sufficiency t is set.[27] This is a sensible thing to say about epistemic justification, but it is also a consequence of SCR (see Appendix 2 for proof) and hence of GCR.

The equi-maximality requirement reduces the candidates for $W(h, e \mid b)$ greatly. Only a handful of measures of incremental confirmation proposed in the literature satisfy this requirement. Crupi, Tentori and Gonzalez (2007) point out that we can obtain an equi-maximal measure of incremental confirmation by "normalizing" any measure of incremental confirmation, i.e. by setting the equi-neutral point at zero and then divide the measure by the maximal value that the measure can take.[28] However, they also demonstrate that many known measures of incremental confirmation, such as $R(h, e \mid b)$ and $D(h, e \mid b)$ mentioned earlier, turn into the same measure $Z^{+}(h, e \mid b)$ when they are "normalized":[29]

$$R(h,e \mid b) = \frac{P(h \mid e \wedge b)}{P(h \mid b)}$$

$$D(h,e \mid b) = P(h \mid e \wedge b) - P(h \mid b)$$

$$Z^{+}(h,e \mid b) = \frac{P(h \mid e \wedge b) - P(h \mid b)}{1 - P(h \mid b)}$$

So, $Z^{+}(h, e \mid b)$ is both equi-neutral and equi-maximal. $Z^{+}(h, e \mid b)$ is also an increasing function of $P(h \mid e \wedge b)$ and a degreasing function of $P(h \mid b)$. Unfortunately, $Z^{+}(h, e \mid b)$ fails to satisfy SCR (and hence GCR) even for $n = 2$, i.e. even when the conjunction in SCR has only two conjuncts. This is easy to see by an example. Suppose h_1 and h_2 are probabilistically independent of each other both on condition of b and on condition of $b \wedge e$. Suppose also that $P(h_1 \mid b) = P(h_2 \mid b) = 0.2$ and $P(h_1 \mid e \wedge b) = P(h_2 \mid e \wedge b) = 0.9$. It follows from these suppositions that $P(h_1 \wedge h_2 \mid b) = 0.2 \times 0.2 = 0.04$ and $P(h_1 \wedge h_2 \mid e \wedge b) = 0.9 \times 0.9 = 0.81$. We can calculate $Z^{+}(h_1, e \mid b)$, $Z^{+}(h_2, e \mid b)$, and $Z^{+}(h_1 \wedge h_2, e \mid b)$, as follows:

$$Z^{+}(h_1,e \mid b) = Z^{+}(h_2,e \mid b) = \frac{0.9 - 0.2}{1 - 0.2} = 0.875$$

$$Z^{+}(h_1 \wedge h_2,e \mid b) = \frac{0.81 - 0.04}{1 - 0.04} = 0.8020833$$

So, even though the two probabilistically independent conjuncts have the same Z^+ value, their conjunction does not, which is in violation of SCR (and hence of GCR).

Fortunately, our problem is solved by "normalizing" the log-ratio measure of incremental confirmation $I(h, e \mid b)$ to obtain $J(h, e \mid b)$ below:

$$\begin{aligned} I(h,e \mid b) &= \log \frac{P(h \mid e \wedge b)}{P(h \mid b)} \\ &= \log P(h \mid e \wedge b) - \log P(h \mid b) \end{aligned}$$

$$\begin{aligned} J(h,e \mid b) &= \frac{\log P(h \mid e \wedge b) - \log P(h \mid b)}{\log 1 - \log P(h \mid b)} \\ &= \frac{\log P(h \mid e \wedge b) - \log P(h \mid b)}{-\log P(h \mid b)} \end{aligned}$$

Since $J(h, e|b)$ satisfies GCR (see Appendix 3 for proof), it is an adequate measure of epistemic worthiness in that it is an increasing function of $P(h \mid e \wedge b)$ and a decreasing function of $P(h \mid b)$, and satisfies GCR.

I want to make two technical points here. First, the choice of the log base does not matter because $[\log_k a]/[\log_k b] = \log_b a$.[30] We can rewrite $J(h, e \mid b)$ with no reference to k as follows:

$$\begin{aligned} J(h,e \mid b) &= \frac{\log_k P(h \mid e \wedge b) - \log_k P(h \mid b)}{-\log_k P(h \mid b)} \\ &= 1 - \frac{\log_k P(h \mid e \wedge b)}{\log_k P(h \mid b)} \\ &= 1 - \log_{P(h|b)} P(h \mid e \wedge b) \end{aligned}$$

The log base k that appears both in the numerator and the denominator of $J(h, e \mid b)$ cancels out and disappears, making no difference in the value of $J(h, e \mid b)$.[31] Second, $J(h, e \mid b)$ is undefined in cases where $P(b) = 0$, $P(e \wedge b) = 0$, $P(h \mid e \wedge b) = 0$, $P(h \mid b) = 0$ or $P(h \mid b) = 1$. The reason in the first two cases is that $P(b) = 0$ and $P(e \wedge b) = 0$ make $P(h \mid b)$ and $P(h \mid e \wedge b)$ undefined, respectively. In both cases, the body of beliefs b is in need of contraction before evaluating any hypothesis. I will assume that $P(b) > 0$ and $P(e \wedge b) > 0$ when the epistemic worthiness of h is measured. In the next two cases, $P(h \mid e \wedge b) = 0$ and $P(h \mid b) = 0$ make $\log P(h \mid e \wedge b)$ and $\log P(h \mid b)$ undefined, respectively. Since h is definitely false in these cases, we can immediately dismiss h without measuring its epistemic worthiness by $J(h, e \mid b)$. I will assume that $P(h \mid e \wedge b) > 0$ and $P(h \mid b) > 0$ when the epistemic worthiness of h is measured.

The last case, where $P(h \mid b) = 1$, is more interesting. $J(h, e \mid b)$ is undefined in this case because $P(h \mid b) = 1$ makes the denominator $-\log P(h \mid b)$ of $J(h, e \mid b)$ zero. Note that if $P(h \mid b) = 1$, then h carries no information at all, but $P(h \mid b) = 1$ also makes $P(h \mid e \wedge b) = 1$ so that h is completely safe. The reason is that when $P(h \mid b) = 1$, h is already implicitly contained in b.[32] So, adding h into b is redundant, which I call the *analytic expansion* of b to distinguish it from the *synthetic expansion* of b in which h adds new information to b. Analytic expansion can be of some practical use when h is easier to understand and easier to apply than those statements in b that establish the truth of h. Since there is no compelling reason to prohibit the redundant but sometimes useful addition, and since the subject may derive logical consequences of b anyways, I consider analytic expansion to be epistemically justified. So, the evaluator has justification for adding h to b under two conditions: when $J(h, e \mid b) \geq t$ (synthetic expansion), and when $P(h \mid b) = 1$ (analytic expansion). I will return to this subject of analytic expansion when I discuss epistemic closure in the next chapter.

Once we find a measure of epistemic worthiness, the next natural question is whether it is the only measure of epistemic worthiness. It is not. We can construct many measures of epistemic worthiness that meet the three conditions, (1) $W(h, e \mid b)$ is an increasing function of $P(h \mid e \wedge b)$, (2) $W(h, e \mid b)$ is a decreasing function of $P(h \mid b)$ and (3) $W(h, e \mid b)$ satisfies GCR. Some of them differ from $J(h, e \mid b)$ in an interesting way. For example, $J(h, e \mid b)$ has the infinite range $(-\infty, 1]$, while Atkinson's (2012) measure $J'(h, e \mid b) = 2^{J(h, e \mid b)} - 1$ has the finite range $[-1, 1]$.[33] However, it also turns out that all measures of epistemic worthiness are ordinally equivalent to each other, and thus to $J(h, e \mid b)$ (see Appendix 4 for proof).[34] In other words, $J(h, e \mid b)$ is the unique measure of epistemic worthiness *up to ordinal equivalence*.

The measure $J(h, e \mid b)$ may look complicated, but it has a simple and intuitive meaning in the language of information. As noted earlier, the amount of information (relative to the body of beliefs b) that h carries is $I(h \mid b) = -\log P(h \mid b)$, according to the standard mathematical measure of information. $I(h \mid b)$ is commonly referred to as "self-information" because it is the amount of information on h that is gained when it becomes certain that h is true. Meanwhile, the amount of information on h (relative to the body of beliefs b) that is gained when it becomes certain that e is true is called "pointwise mutual information" and is defined as follows: $I(h, e \mid b) = \log P(h \mid e \wedge b) - \log P(h \mid b)$.[35] To see its intuitive meaning, suppose $P(h \mid b)$ is 1/8 and the evidence e raises its probability to $P(h \mid e \wedge b) = 1/2$. Then, the amount of pointwise mutual information in bits (i.e. with log base 2) is $I(h, e \mid b) = \log_2 1/2 - \log_2 1/8 = \log_2 2^{-1} - \log_2 2^{-3} = 2$. This means that we gain 2 bits of information on h when we obtain the evidence e. Self-information $I(h \mid b) = -\log P(h \mid b)$ can be rewritten in terms of pointwise mutual information as $I(h, h \mid b) = \log P(h \mid h \wedge b) - \log P(h \mid b) = \log 1 - \log P(h \mid b) = -\log P(h \mid b)$. It is the amount of pointwise mutual information on h we gain in case the evidence e obtained is h itself.

Note here that the numerator of $J(h, e \mid b)$ is the pointwise mutual information $I(h, e) = \log P(h \mid e \wedge b) - \log P(h \mid b)$, while the denominator of $J(h, e \mid b)$ is the self-information $I(h, h \mid b) = -\log \mathrm{P}(h \mid b)$. So, $J(h, e \mid b)$ turns out to be the ratio of the pointwise mutual information to the self-information:

$$J(h, e \mid b) = \frac{I(h, e \mid b)}{I(h, h \mid b)}$$

This expression allows us to interpret the degree of epistemic worthiness $J(h, e \mid b)$ in a natural way. Self-information $I(h, h \mid b)$ is the amount of information that we *register* when we add h to b. We may call it "the registered information". Meanwhile pointwise mutual information $I(h, e \mid b)$ is the amount of information on h we gain from the evidence e. We may call it "the earned information". If we use this terminology, the degree of epistemic worthiness $J(h, e \mid b)$ is the ratio of the earned information to the registered information. The higher the ratio is, the more epistemically worthy it is to add h to b.

We can also interpret $J(h, e \mid b)$ in terms of the unearned information by rewriting $J(h, e \mid b)$ as follows:

$$\begin{aligned} J(h, e \mid b) &= \frac{\log P(h \mid e \wedge b) - \log P(h \mid b)}{-\log P(h \mid b)} \\ &= 1 - \frac{-\log P(h \mid e \wedge b)}{-\log P(h \mid b)} \end{aligned}$$

We can regard $-\log P(h \mid e \wedge b)$ as the amount of unearned information because we obtain it by subtracting the earned information, $\log P(h \mid e \wedge b) - \log P(h \mid b)$, from the registered information, $-\log P(h \mid b)$. So, the second term (the subtrahend) is the ratio of the unearned information to the registered information. The higher this ratio is, the less epistemically worthy h is. This makes good sense if justification of the statement based on its epistemic worthiness is to serve the dual goal of cognition—to increase true beliefs and avoid false beliefs.

The alternative expression also reveals that the threshold of sufficient epistemic worthiness t for acceptance must be positive. Otherwise, a hypothesis for which none of the registered information is earned could be considered sufficiently epistemically worthy. Of course, the threshold is much greater than zero in communities with substantial epistemic resources. So, we obtain the following dual-component format of epistemic evaluation for the synthetic expansion of the body of beliefs:[36]

> **The Dual-Component Format of Epistemic Evaluation:** Given the body of evidence e, there is epistemic justification for the evaluator to endorse the synthetic expansion of the subject's body of beliefs b by the addition

of the statement h if and only if the degree of epistemic worthiness $J(h, e \mid b)$ is no smaller than t, where $t > 0$ is the threshold of sufficient epistemic worthiness for acceptance.

4.5 Comparison with Similar Approaches

I have shown above that measure J is the unique measure (up to ordinal equivalence, hereafter this qualification is not repeated) of epistemic worthiness. This is important when we apply the measure to epistemological questions—for example, when we compare the epistemic worthiness of the natural world hypothesis, the BIV hypothesis, and their disjunction. If there are many measures of epistemic worthiness and the order of epistemic worthiness among the hypotheses is dependent on the choice of a measure, the answer to the question by any particular measure is not definitive. The answer by measure J is definitive because it is the unique measure of epistemic worthiness. There are, however, some proposals in the literature that resemble the dual-component format of epistemic evaluation developed in this chapter. Some people may wonder whether the answer by measure J—though definitive in the dual-component format—may not be definitive unless these proposals are either rejected or shown to agree with the evaluation by the dual-component format. I cannot examine all such proposals here, but I want to discuss two of them that resemble the dual-component format.

One of them is the truthlikeness approach, which revives and develops Popper's concept of verisimilitude.[37] To put the idea of truthlikeness in a simple way, if the language we use on the given subject has finitely many atomic statements, we can measure the informativeness of a theory on the subject by the ratio of the true atomic statements captured by the theory. We can also measure the ratio of the false atomic statements contained in the theory. The degree of truthlikeness is then determined by subtracting the latter from the former. There are many ways of making this basic idea precise, but the truthlikeness approach in general has some features in common that are incompatible with the dual-component format developed in this chapter. Most importantly, it is possible in the truthlikeness approach that a false theory (a theory that contains a false statement) is closer to the (whole) truth than a true theory (a theory that contains no false statement) if the former is much more informative than the latter. For example, the false theory may contain numerous atomic statements and only one of them is false so that the degree of truthlikeness is very high even after subtraction, while the true theory may contain only a few atomic statements so that the degree of truthlikeness is very low even though there is no subtraction.

The idea of truthlikeness is similar to the dual-component format, which takes into account both safety and informativeness, but the dual-component format never endorses a statement that is definitely false or withholds endorsement for a statement that is definitely true. The two approaches are incompatible. However, there is actually no need for rejecting one approach or the other because they are developed for different purposes. The truthlikeness approach is developed for the purpose of measuring the progress of science. For example, the measure of truthlikeness is applied to two theories on the same subject from different times in the history of science. Even if they are both false by the contemporary standard, we may still say science made progress if the more recent one is closer to the (whole) truth than is the older one. As a result of its aim of assessing progress in science, the truthlikeness approach has features that are not suitable for evaluating the epistemic worthiness of a statement for inclusion into the body of beliefs.[38]

First, theories to evaluate are the prevailing theories from the past, and it is not the aim of the truthlikeness approach to modify them for improvement. Second, each theory to evaluate is a collection of statements, and the concept of truthlikeness is not applicable to an individual statement in it. This is in contrast with the concept of epistemic worthiness, which is applicable to an individual statement. Also, when we apply the concept of epistemic worthiness to a collection of statements, we can modify the collection for improvement. For example, if some statement in the collection is definitely false in light of some evidence, we simply drop it and evaluate the reduced package, instead of evaluating the original package. In general, there is no need to measure epistemic worthiness if a statement is definitely true or definitely false because the former is clearly acceptable while the latter is clearly unacceptable. The measure of epistemic worthiness is intended for a statement that may or may not be true in light of evidence—to determine whether it is worthy of acceptance based on its epistemic probability and informativeness. Since the dual-component format and the truthlikeness approach have different objectives, there is no need to reject one approach or the other. For the purpose of evaluating the epistemic worthiness of a statement that may or may not be true, we should use the dual-component format.[39]

The other approach I want to consider for comparison is decision theory, which combines probabilities and utilities to calculate the expected utility of an action. The dual-component format bears some resemblance to it: Just as it is rational by decision theory to gamble on a low-probability outcome if the potential gain is sufficiently high, it is rational by the dual-component format to accept a low-probability statement if the potential gain in truth (measured by the amount of information) is sufficiently high. Of course, at stake in epistemic evaluation is the alethic gain, but it looks like a species of utility. We may therefore apply the general framework of decision theory to evaluate belief by its expected utility (the weighted average of utilities).[40] The resulting evaluation is at odds with the dual-component format. The measure of epistemic worthiness balances safety and informativeness, but not in the way decision theory does—the degree of epistemic worthiness is not the weighted average of potential gains in truth. Is decision theory a legitimate alternative to the dual-component format of epistemic evaluation?

Setting aside the types of utility, it is important that the expected utility is a measure of the *instrumental* value of an action. An action is deemed valuable if it produces good results on average. If we apply decision theory to the evaluation of belief, then we also need to consider the instrumental value of belief. In contrast, the dual-component format of epistemic evaluation disregards the instrumental value. Informativeness, which measures the potential gain in truth, is the amount of information the belief itself carries. The distinction drawn by Firth (1998) between "intrinsic epistemic merit" and "instrumental epistemic merit" is helpful here. Firth considers a case of a researcher who believes against evidence that she will recover from illness, which improves her chances of discovering important truths. Her groundless belief in recovery has little intrinsic epistemic merit, but it has considerable instrumental epistemic merit. To use Firth's terms, epistemic worthiness measures the intrinsic epistemic merit of a statement in disregard of its instrumental merit. It has a narrower objective than does the decision-theoretic approach to the evaluation of beliefs.

The question arises at this point whether epistemic evaluation should take into account the instrumental epistemic merits of beliefs. In the case of the researcher, for example, should the epistemologist consider the value of important truths she may discover as a result of believing in her recovery from illness? In recent articles Berker (2013a, 2013b) takes the position that instrumental epistemic merits are irrelevant to epistemic evaluation.[41] He uses this point against teleological theories in epistemology on the ground that they sanction beliefs with little intrinsic epistemic merit in "Firth-style cases". The scope of Berker's objection is unclear because some theories in epistemology that Berker considers teleological, such as process reliabilism (Goldman 1979), set aside instrumental epistemic merits in their epistemic evaluation.[42] The more interesting question, however, is whether we have good reason to take into account the instrumental epistemic merit of belief in meliorative epistemology. It looks like the answer is yes because beliefs with significant instrumental epistemic merits help us achieve the alethic goals of cognition.

There is some complication to note. The evaluator in meliorative *social* epistemology endorses a statement to all members of the epistemic community without knowing the specific circumstances of a particular member that determine the instrumental epistemic merits of accepting the statement. It seems impossible to incorporate instrumental epistemic merits into meliorative social epistemology. However, statements evaluated in meliorative social epistemology include epistemic guidelines. The evaluator may endorse some epistemic guideline based on general principles of psychology to promote beliefs that tend to increase instrumental epistemic merits—perhaps a guideline that inflates the estimate of one's own competence in research.

Despite their contribution to the alethic goals, I set aside instrumental epistemic merits and stand by the dual-component format. The inclusion of instrumental epistemic merits blurs the distinction between epistemic and pragmatic reasons for belief. It is easy to see that Firth-style cases are a variant of more common cases where the instrumental *pragmatic* merits of the belief offset the lack of evidence in its support. For example, a patient who believes against evidence that she will recover from illness may live longer and more happily because of her groundless belief. The instrumental merits involved here is pragmatic, but her groundless belief is justified in the same way, essentially, by the good consequences the belief produces. Note further that the pragmatic justification of belief is a variant of more common cases where an *action* is justified by the good consequences it produces. For example, a patient may have good reason to do yoga, adopt a dog, etc. if it makes her live longer and more happily. In short, Firth-style cases have much in common with the pragmatic evaluation of an action by its consequences.

Once we include instrumental epistemic merits in the evaluation of belief, it is more sensible to regard epistemic evaluation as a special case of the pragmatic evaluation of an action, where the action evaluated happens to be the action of believing, and the relevant positive consequence happens to be the discovery of truth. It then becomes sensible to evaluate the action of believing in comparison with other courses of actions, and by the balance of all consequences—both alethic and non-alethic.[43] The upshot is that epistemic evaluation is no longer an independent category of evaluation. Some people may welcome this perspective.[44] However, blurring the distinction between the epistemic and pragmatic justification makes the evaluation of belief inter-personally unstable because different people value different consequences. Further, even if people agree on what consequences are valuable, the evaluation of belief becomes cross-temporally unstable because information with no bearing on the truth of the belief can change its value at any point. The problem of instability is exacerbated when the belief in question is combined with other beliefs one holds, and further with beliefs held by others in the community.[45] It becomes impossible to tell whether a particular belief will have good overall consequences in the long run. By setting aside consequences of the belief and focusing on its intrinsic epistemic merits, we can keep epistemic evaluation independent and largely stable.

Some people may point out that even if we make the epistemic evaluation of a particular belief stable by disregarding its instrumental merits, the independence of epistemic evaluation is lost quickly when the belief is combined with other beliefs, some of which are accepted for pragmatic reasons. There are clear cases where pragmatic considerations override epistemic evaluation. James (1979) provides a classic example of a mountain climber in a precarious position whose only way out is a dangerous leap, but who has no evidence that he will succeed. Given human psychology, firmly believing that one will make it to the other side raises the chance of success. In James's words "*faith creates its own verification*"

(p. 80, original italics). It is hard to deny in such cases that the pragmatic circumstances justify the formation of belief by faith in the absence of evidence or even against evidence. We may call them "faith-based beliefs", and once some faith-based beliefs are formed for good pragmatic reasons, they will be combined with other beliefs one holds, and further with beliefs held by others in the community. The result is a network of beliefs justified by the mixture of epistemic and pragmatic reasons. There is, in other words, no way of avoiding the messy entanglement of the epistemic and pragmatic evaluation in the end.

There is actually an easy way of avoiding entanglement: We keep faith-based beliefs temporal and personal. They should not be part of the body of beliefs we use as the basis of diverse deliberation. We should incorporate statements into the body of beliefs only if they are epistemically worthy in light of its content and evidence. The body of beliefs then serves as the mostly stable basis of diverse deliberations. If for some reason we need to keep some faith-based beliefs over time, they should be kept apart from the body of evidence-based beliefs, so that they should not be combined with other beliefs we hold. If for some reason we need to share faith-based beliefs with others, we should clarify that they are based on faith, and not evidence, so that they will not be combined with evidence-based beliefs in their mind.

To summarize, the main point of the chapter is to replace the Lockean format of epistemic evaluation by the dual-component format of epistemic evaluation. The Lockean format is rejected because it allows the backdoor endorsement of an unacceptable hypothesis. The dual-component format, which blocks backdoor endorsement, takes into account both the safety and informativeness of the statement. I then introduced the specific measure *J* of epistemic worthiness, and showed that it is the unique measure (up to ordinal equivalence) that meets the formal constraints required for blocking backdoor endorsement. The uniqueness of the measure is important because when the epistemic worthiness of various hypotheses—the natural world hypothesis, the BIV hypothesis and their disjunction—is compared in the next chapter, the answer given by the measure *J* will be definitive. To reinforce this point, I discussed two approaches that resemble the dual-component format to show that they are not appropriate for epistemic evaluation of a statement for inclusion in the body of beliefs. The dual-component format is the right format of qualitative epistemic evaluation to determine whether we should accept the natural world hypothesis in the face of Cartesian skepticism.

Notes

1 To recall, the reflective sensory evidence *e* describes the content of the subject's sensory experience, e.g. it appears to me visually that here is a hand.

2 See Ramsey (1931), de Finetti (1937), and Jeffrey (1983) among many others. For the opposing "objectivist" strain of Bayesianism, see Jeffreys (1939), Jaynes (2003), Williamson (2010) among many others.

3 The modality here is *can* in the sense of permission. A positive qualitative epistemic evaluation of a statement gives us a permission to accept it, but we are not obligated to accept it. Even if we recognize some statement to be true, we may not register it in the form of a belief if it is uninteresting and unlikely to impact our life. We are not obligated, for example, to form the belief that there was rain in Vladivostok yesterday even if a reliable weather report says so. The point is not that we may withhold epistemic judgment in the face of strong evidence, but that we need not register it in the form of a belief because registration comes with some cost, e.g. it uses up some memory space. It is therefore unwise to register all statements with high epistemic standings in the form of beliefs. Harman (1989) calls it "the principle of clutter avoidance".

4 Or, alternatively, from $P(e \mid h \wedge b)$, $P(e \mid \neg h \wedge b)$ and $P(h \mid b)$.

5 I will discuss statistical learning theory in Chapter 6 and propose a format that evaluates probability assignments in light of the body of evidence.

6 The term is due to Foley (2009), though Foley uses it to characterize rational belief instead of epistemic justification.

7 Being a member of the community herself, the evaluator accepts all statements she endorses to the community, but the converse does not hold. The evaluator may not endorse to the subject some statements she accepts because the principle of clutter avoidance (mentioned in endnote 3) also applies to endorsement. For example, the evaluator would not endorse a huge amount of experimental data for the members of the community to register in the form of beliefs. To avoid clutter, the evaluator would endorse only those statements that are interesting to the members of the community and likely to impact their life. Those statements the evaluator accepts but does not endorse to the community are part of the evidence available to the evaluator.

8 To be more precise, the issue is closure under *known* logical entailment. I will assume that the logical entailment in question is known to both the evaluator and the subject in the case.

9 The lottery paradox was introduced by Kyburg (1961). See Wheeler (2007) for an extensive review of the literature on the lottery paradox.

10 To put the point informally, the evaluator can endorse *each* of the premises, but not *all* of the premises.

11 The solution makes the acceptance of a particular premise sensitive to the order of evaluation. As just mentioned, once the epistemic subject adds $p_1, p_2, p_3, \ldots, p_{n-1}$ to her body of beliefs, she can no longer add p_n to the expanded body of beliefs. However, if the order of evaluation is reversed, so that p_n is evaluated before the other premises, then she can add p_n to her body of beliefs. This may make some people feel uneasy, but it is unclear whether this leads to a serious objection to the solution because the probabilistic profiles of the premises are the same. I will discuss cases in Chapter 5 where the probabilistic profiles of the statements to choose from are not the same.

12 This feature makes the case similar to the preface paradox (Makinson 1965). The similarity is that accepting all of them is not logically inconsistent, but each statement in the preface paradox is not probabilistically independent of each other. Some have used the preface paradox to argue against the Lockean format and motivate the notion of truthlikeness (Cevolani 2017; Cevolani and Schurz 2017). I will comment on the truthlikeness approach in Section 5.

13 The probability distribution does not violate the rules of the probability calculus. In particular, the condition (4) $P(d_1 \wedge \ldots \wedge d_n \mid \neg h \wedge e \wedge b) = 0$ does not make $P(p_1 \vee \ldots \vee p_n \mid e \wedge b)$ exceed one because $P(p_1 \vee \ldots \vee p_n \mid e \wedge b) = 1 - P(\neg p_1 \wedge \ldots \wedge \neg p_n \mid e \wedge b)$, where the subtrahend $P(\neg p_1 \wedge \ldots \wedge \neg p_n \mid e \wedge b)$ cannot be negative.

14 As explained in Chapter 1, meliorative epistemology seeks the best format of epistemic evaluation to achieve the alethic goals of cognition under the constraints

of available epistemic resources, but the alethic goals of cognition are not fixed forever. We may give up part of the original goals if it turns out to be unachievable under the given resource constraints.

15 I called it a measure of "epistemic justification" in the past (Shogenji 2012), but the standard Bayesian measure is also called a measure of "epistemic justification" in the literature. I am using a different term in this book to avoid unnecessary confusion and unproductive disputes over terminology.

16 For example, if p_1 logically entails p_2, then the second conjunct p_2 of the conjunction $p_1 \wedge p_2$ is redundant because $p_1 \wedge p_2$ if and only if p_1. If p_2 is redundant, accepting $p_1 \wedge p_2$ is no riskier than accepting p_1.

17 In cases where Popper's pessimism is appropriate and the probability that a certain statement is true is zero or near zero, it is more appropriate to apply the relative-divergence format of epistemic evaluation (to be introduced in Chapter 6), which allows us to differentiate false theories by their (estimated) degrees of inaccuracy. Some philosophers of science develop Popper's idea in a different way by the notion of "truthlikeness". I will discuss the truthlikeness approach in Section 5 of this chapter.

18 See Huber (2008a, 2008b) for a two-factor approach to the formal assessment of scientific theories. Huber calls the two factors "plausibility" and "informativeness". See Kaplan (1981a, 1981b, 1996, Chapter 4) for a two-factor account of "rational acceptance" in epistemology in general. Kaplan calls the two factors "truth" and "comprehensiveness". My account below goes beyond theirs in that it proposes and defends the precise way the two factors should be balanced.

19 The log base k must be greater than 1 to make $I(h) = -\log_k P(h)$ a decreasing function of $P(h)$.

20 See Maher (1993, Section 9.5.1) for a critical examination of various measures of information.

21 Levi (1980) proposes a third alternative, according to which "informational utility" is determined by the probability function M that reflects the individual's *values*. M is a probability function in the formal sense of satisfying the axioms of the probability calculus, but the value-reflecting function M does not assign "probabilities" in the ordinary sense, i.e. the probabilities that the statements are *true*. I do not pursue this proposal here since Levi does not provide detailed accounts of M, leaving its nature somewhat mysterious.

22 Two measures $M(h, e \mid b)$ and $M^*(h, e \mid b)$ are ordinally equivalent to each other if and only if for any two ordered triples $<h_1, e_1, b_1>$ and $<h_2, e_2, b_2>$, $M(h_1, e_1 \mid b_1) <$ (or =, >) $M(h_2, e_2 \mid b_2)$ if and only if $M^*(h_1, e_1 \mid b_1) <$ (or =, > $M^*(h_2, e_2 \mid b_2)$, respectively.

23 Unlike the resource constraints that are determined by the epistemic resources available to the community, formal constraints are grounded in the alethic goals of cognition. An adequate measure of epistemic worthiness must satisfy the two formal constraints (the backdoor constraint and the inverse backdoor constraint) because the subject cannot achieve the alethic goals of cognition if the same statement is both endorsed and not endorsed.

24 The existence of the neutral value may prompt the suggestion that the threshold of sufficiency t must be positive, so that there is no justification for adding h to b unless h has a positive degree of epistemic worthiness. If we choose to restrict the range of the threshold t in this way, then GCR would be of the form "for some threshold $t > 0$" instead of "for some threshold t". The restricted version of GCR entails the restricted version of SCR whose condition is "if all the conjuncts have the same *positive* degree of epistemic worthiness".

25 See Fitelson (1999), Crupi, Tentori and Gonzalez (2007), Roche and Shogenji (2014) for a growing list of measures of incremental confirmation proposed in the literature.

26 For example, some people suggest that it is desirable for a measure of incremental confirmation C to have the property of "Hypothesis Symmetry" (Fitelson 2002). It means that the degree to which h is confirmed by e is the same as the degree to which its negation $\neg h$ is disconfirmed by e, i.e. $C(h, e \mid b) = -C(\neg h, e \mid b)$. The property may be desirable for a measure of incremental confirmation, but there is no reason to think it is desirable for a measure of epistemic worthiness.

27 For now I set aside the case where the statement h is already certain on condition of the body of beliefs b alone, so that it is not the case that $P(h \mid e \wedge b) = P(h \mid b) = 1$ to make $W(h, e \mid b)$ *both* neutral and maximal. I will address this case shortly, after formulating the specific measure J of epistemic worthiness.

28 Crupi, Tentori and Gonzalez (2007) use the same procedure to make the measure of incremental confirmation *equi-minimal*, i.e. make the value constant (set at minus one) when $P(h \mid e \wedge b) = 0$, regardless of $P(h \mid b)$. This is because they want the degree of incremental confirmation to be a generalization of logical relation. I set aside this aspect of their proposal.

29 Z^{+} is the positive half of Crupi, Tentori and Gonzalez's (2007) measure Z of incremental confirmation. I made the body of beliefs (the "background beliefs") b explicit in all of them, though in many cases measures of incremental confirmation are stated without explicitly mentioning it.

30 Let $\log_b a = x$ so that $a = b^x$. It follows that $\log_k a = \log_k b^x = x \log_k b$, and thus $\log_b a = x = [\log_k a] / [\log_k b]$.

31 It is noted earlier that the log base k must be greater than one, so that $I(h \mid b) = -\log_k P(h \mid b)$ is a decreasing function of $P(h \mid b)$. However, the log base $P(h \mid b)$ used for eliminating k is less than one. As a result, $-\log_{P(h \mid b)} P(h \mid e \wedge b)$ is an increasing function of $P(h \mid e \wedge b)$. This is not a problem and is actually consistent with the condition that the epistemic worthiness of h is an increasing function of its safety.

32 Recall the assumption that the probability distribution is regular, so that $P(h \mid b) = 1$ only if b logically entails e.

33 I add the background beliefs b here, which is suppressed in Atkinson's formulation.

34 See also Atkinson (2012) for an illuminating alternative proof.

35 $I(x, y \mid b)$ is called "pointwise" mutual information because it is the amount of information that the particular statement y provides on the particular statement x. Their weighted average $I(X, Y \mid \mathrm{b})$ is called "mutual information", where X and Y are partitions instead of particular statements. $I(x, y \mid b)$ is called pointwise "mutual" information because it follows from the definition that $I(x, y \mid b) = I(y, x \mid b)$.

36 The format will be qualified to account for the possibility of supersedure in the next chapter.

37 The original account of verisimilitude proposed by Popper (1963, 1972) turned out to have serious technical difficulties (Miller 1974; Tichý 1974). Recent advocates of truthlikeness (Niiniluoto 1987, 1999, 2013; Kuipers 2000, among others) explore various versions that circumvent the technical difficulties.

38 Conversely, the dual-component format is not appropriate for measuring the progress of science. I will introduce the relative-divergence format (Chapter 6) that can be used for that purpose, though it is developed primarily for the purpose of model selection.

39 Some truthlikeness advocates propose to use *estimated* truthlikeness for choosing a theory (Niiniluoto 1987, Ch. 6) beyond assessing the progress of science, but the proposal still retains some of the features suitable for measuring the

progress of science. For example, a theory can have a high degree of estimated truthlikeness even if it contains a statement that is definitely false.

40 Isaac Levi develops this idea in his aptly titled book *Gambling with Truth* (1967).

41 The reason for Berker's position is the pre-theoretical judgment that the subject has no epistemic justification for her belief even if it has significant instrumental epistemic merits.

42 Goldman (2015) makes this point for process reliabilism in response to Berker.

43 The action of believing is usually compatible with other actions. When the actions that are considered are compatible, the real options to choose from are the Boolean combinations of the actions.

44 See, for example, the suggestion of pragmatic encroachment in epistemology (Fantl and McGrath 2002, 2007, 2009).

45 This is one of the points Clifford (1999) makes in his argument against believing anything on insufficient evidence, i.e. even if some belief held on insufficient evidence causes no direct harm on its own, it can still cause serious harm when it is combined with other beliefs one holds, and further with beliefs held by others in the community.

Appendices

1. Equi-Neutrality

Suppose $W(h, e \mid b)$, which is of the form $F(P(h \mid e \wedge b), P(h \mid b))$, is a continuous function, and $W(h, e \mid b)$ satisfies SCR. Then, for any two pairs $<h_i, e_i>$ and $<h_j, e_j>$, if $P(h_i \mid e_i \wedge b) = P(h_i \mid b)$ and $P(h_j \mid e_j \wedge b) = P(h_j \mid b)$, then $W(h_i, e_i \mid b) = W(h_j, e_j \mid b)$.

Proof

Let $\log_{P(h_i|b)} P(h_j \mid b) = r$, so that $[P(h_i \mid b)]^r = P(h_j \mid b)$. It follows from $0 < P(h_i \mid b) < 1$ and $0 < P(h_j \mid b) < 1$ that $r > 0$. Since $W(h, e \mid b)$ is a continuous function, it suffices to show that the claim holds for any two pairs $<h_i, e_i>$ and $<h_j, e_j>$ such that $[P(h_i \mid b)]^q = P(h_j \mid b)$, where q is a positive rational number. Let $<m, n>$ be the smallest pair of positive integers such that $n/m = q$, and thus $[P(h_i \mid b)]^n = [P(h_j \mid b)]^m$. Choose probabilistically independent (both on condition of b and on condition of $b \wedge e_i$) statements $h_1, \ldots, h_n$, and probabilistically independent (both on condition of b and on condition of $b \wedge e_j$) statements $h_{n+1}, \ldots, h_{n+m}$ such that:[1]

(i) $[P(h_i \mid b)]^n = [P(h_j \mid b)]^m$
(ii) $P(h_i \mid b) = P(h_1 \mid b) = \ldots = P(h_n \mid b)$
(iii) $P(h_j \mid b) = P(h_{n+1} \mid b) = \ldots = P(h_{n+m} \mid b)$
(iv) $P(h_i \mid e_i \wedge b) = P(h_1 \mid e_i \wedge b) = \ldots = P(h_n \mid e_i \wedge b)$
(v) $P(h_j \mid e_j \wedge b) = P(h_{n+1} \mid e_j \wedge b) = \ldots = P(h_{n+m} \mid e_j \wedge b)$

It follows from (ii) and (iv) that $W(h_i, e_i \mid b) = W(h_1, e_i \mid b) = \ldots = W(h_n, e_i \mid b)$. So, by SCR:

$$W(h_i, e_i \mid b) = W(h_1 \wedge \ldots \wedge h_n, e_i \mid b) \tag{1}$$

Similarly, it follows from (iii) and (v) that $W(h_j, e_j \mid b) = W(h_{n+1}, e_j \mid b) = \ldots = W(h_{n+m}, e_j \mid b)$. So, by SCR:

$$W(h_j, e_j \mid b) = W(h_{n+1} \wedge \ldots \wedge h_{n+m}, e_j \mid b) \tag{2}$$

Since $h_1, \ldots, h_n$ are probabilistically independent on condition of b, $P(h_1 \wedge \ldots \wedge h_n \mid b) = [P(h_i \mid b)]^n$ from (ii). Similarly, since $h_{n+1}, \ldots, h_{n+m}$ are probabilistically independent on condition of b, $P(h_{n+1} \wedge \ldots \wedge h_{n+m} \mid b) = [P(h_j \mid b)]^m$ from (iii). So, it follows from (i) that:

$$P(h_1 \wedge \ldots \wedge h_n \mid b) = P(h_{n+1} \wedge \ldots \wedge h_{n+m} \mid b) \tag{3}$$

Since $h_1, \ldots, h_n$ are probabilistically independent on condition of $e_i \wedge b$, $P(h_1 \wedge \ldots \wedge h_n \mid e_i \wedge b) = [P(h_i \mid e_i \wedge b)]^n = [P(h_i \mid b)]^n$ from (iv) and the condition $P(h_i \mid e_i \wedge b) = P(h_i \mid b)$ in the antecedent of the theorem. Similarly, since $h_{n+1}, \ldots, h_{n+m}$ are probabilistically independent on condition of $e_j \wedge b$, $P(h_{n+1} \wedge \ldots \wedge h_{n+m} \mid e_j \wedge b) = [P(h_j \mid e_j \wedge b)]^m = [P(h_j \mid b)]^m$ from (v) and the condition $P(h_j \mid e_j \wedge b) = P(h_j \mid b)$ in the antecedent of the theorem. So, it follows from (i) that:

$$P(h_1 \wedge \ldots \wedge h_n \mid e_i \wedge b) = P(h_{n+1} \wedge \ldots \wedge h_{n+m} \mid e_j \wedge b) \tag{4}$$

From (3) and (4) it follows that:

$$W(h_1 \wedge \ldots \wedge h_n, e_i \mid b) = W(h_{n+1} \wedge \ldots \wedge h_{n+m}, e_j \mid b) \tag{5}$$

From (1), (2) and (5) it follows that $W(h_i, e_i \mid b) = W(h_j, e_j \mid b)$ ■

2. Equi-Maximality

Suppose $W(h, e \mid b)$ is a measure of epistemic worthiness (i.e. it is an increasing function of $P(h \mid e \wedge b)$, a decreasing function of $P(h \mid b)$, and satisfies GCR and hence SCR). Then, for any two pairs $<h_i, e_i>$ and $<h_j, e_j>$, if $P(h_i \mid e_i \wedge b) = P(h_j \mid e_j \wedge b) = 1$, then $W(h_i, e_i \mid b) = W(h_j, e_j \mid b)$.

Proof

Assume without loss of generality that $P(h_i \mid b) \leq P(h_j \mid b)$. It follows from this and from the condition $P(h_i \mid e_i \wedge b) = P(h_j \mid e_j \wedge b)$ in the antecedent of the theorem that:

$$W(h_i, e_i \mid b) \geq W(h_j, e_j \mid b) \tag{1}$$

since $W(h, e \mid b)$ is a decreasing function of $P(h \mid b)$. Choose probabilistically independent (on condition of b) statements $h_1, \ldots, h_n$ such that:

(i) $[P(h_j \mid b)]^n \leq P(h_i \mid b)$
(ii) $P(h_j \mid b) = P(h_1 \mid b) = \ldots = P(h_n \mid b)$

It follows from (ii) and $P(h_1 \mid h_1 \wedge \ldots \wedge h_n \wedge b) = \ldots = P(h_n \mid h_1 \wedge \ldots \wedge h_n \wedge b) = 1$ that $W(h_1, h_1 \wedge \ldots \wedge h_n \mid b) = \ldots = W(h_n, h_1 \wedge \ldots \wedge h_n \mid b)$. But $h_1, \ldots, h_n$ are probabilistically independent on condition of b, and $h_1, \ldots, h_n$ are also trivially probabilistically independent on condition of $h_1 \wedge \ldots \wedge h_n \wedge b$ because $P(h_1 \mid h_1 \wedge \ldots \wedge h_n \wedge b) = \ldots = P(h_n \mid h_1 \wedge \ldots \wedge h_n \wedge b) = 1$. So, by SCR:

$$W(h_1, h_1 \wedge \ldots \wedge h_n \mid b) = W(h_1 \wedge \ldots \wedge h_n, h_1 \wedge \ldots \wedge h_n \mid b) \tag{2}$$

Also, it follows from the condition $P(h_j \mid e_j \wedge b) = 1$ in the antecedent of the theorem and $P(h_1 \mid h_1 \wedge \ldots \wedge h_n \mid b) = 1$ that $P(h_j \mid e_j \wedge b) = P(h_1 \mid h_1 \wedge \ldots \wedge h_n \wedge b)$. Further, $P(h_j \mid b) = P(h_1 \mid b)$ from (ii). So,

$$W(h_j, e_j \mid b) = W(h_1, h_1 \wedge \ldots \wedge h_n \mid b) \tag{3}$$

It follows from (2) and (3) that:

$$W(h_j, e_j \mid b) = W(h_1 \wedge \ldots \wedge h_n, h_1 \wedge \ldots \wedge h_n \mid b) \tag{4}$$

Meanwhile, it follows from (ii) that $P(h_1 \wedge \ldots \wedge h_n \mid b) = [P(h_j \mid b)]^n$ since $h_1, \ldots, h_n$ are probabilistically independent on condition of b. But $[P(h_j \mid b)]^n \leq P(h_i \mid b)$ from (i). So,

$$P(h_1 \wedge \ldots \wedge h_n \mid b) £ P\left(h_i \mid b\right) \tag{5}$$

while it follows from the condition $P(h_i \mid e_i \mid b) = 1$ in the antecedent of the theorem and $P(h_1 \wedge \ldots \wedge h_n \mid h_1 \wedge \ldots \wedge h_n \wedge b) = 1$ that:

$$P(h_1 \wedge \ldots \wedge h_n \mid h_1 \wedge \ldots \wedge h_n \wedge b) = P(h_i \mid e_i \wedge b) \tag{6}$$

It follows from (5) and (6) that:

$$W(h_1 \wedge \ldots \wedge h_n \mid h_1 \wedge \ldots \wedge h_n \mid b) \geq W(h_i, e_i \mid b) \tag{7}$$

since $W(h, e \mid b)$ is a decreasing function of $P(h \mid b)$. It follows from (4) and (7) that:

$$W(h_j, e_j \mid b) \geq W(h_i, e_i \mid b) \tag{8}$$

It follows from (1) and (8) that $W(h_i, e_i \mid b) = W(h_j, e_j \mid b)$ ■

3. Satisfaction of the General Conjunction Requirement

Suppose $h_1, \ldots, h_n$ are probabilistically independent (both on condition of b and on condition of $b \wedge e$) and $P(h_1 \mid b), \ldots, P(h_n \mid b) < 1$. Then, (i) if $J(h_1, e \mid b), \ldots, J(h_n, e \mid b) \geq t$, then $J(h_1 \wedge \ldots \wedge h_n, e \mid b) \geq t$; (ii) if $J(h_1, e \mid b), \ldots, J(h_n, e \mid b) < t$, then $J(h_1 \wedge \ldots \wedge h_n, e \mid b) < t$.

Proof

$$J(h_1 \wedge \ldots \wedge h_n, e \mid b) = \frac{\log P(h_1 \wedge \ldots \wedge h_n \mid e \wedge b) - \log P(h_1 \wedge \ldots \wedge h_n \mid b)}{-\log P(h_1 \wedge \ldots \wedge h_n \mid b)}$$

$$= \frac{\log \prod_{i=1}^{n} P(h_i \mid e \wedge b) - \log \prod_{i=1}^{n} P(h_i \mid b)}{-\log \prod_{i=1}^{n} P(h_i \mid b)}$$

$$= \frac{\sum_{i=1}^{n} \log P(h_i \mid e \wedge b) - \sum_{i=1}^{n} \log P(h_i \mid b)}{-\sum_{i=1}^{n} \log P(h_i \mid b)}$$

$$= \frac{\sum_{i=1}^{n} [\log P(h_i \mid e \wedge b) - \log P(h_i \mid b)]}{\sum_{i=1}^{n} -\log P(h_i \mid b)}$$

[from independence] (1)

(i) Suppose $J(h_1, e \mid b), \ldots, J(h_n, e \mid b) \geq t$. Then, for any $i = 1, \ldots, n$, there is some $\alpha_i \geq 0$ such that:

$$J(h_i, e \mid b) = \frac{\log P(h_i \mid e \wedge b) - \log P(h_i \mid b)}{-\log P(h_i \mid b)}$$

$$= t + a_i$$

It follows that:

$$\log P(h_i \mid e \wedge b) - \log P(h_i \mid b) = (t + a_i)[-\log P(h_i \mid b)]$$

By plugging this into (1) above, we obtain:

$$J(h_1 \wedge \ldots \wedge h_n, e \mid b) = \frac{\sum_{i=1}^{n} (t + \alpha_i)[-\log P(h_i \mid b)]}{\sum_{i=1}^{n} -\log P(h_i \mid b)}$$

$$= \frac{t \sum_{i=1}^{n} -\log P(h_i \mid b) + \sum_{i=1}^{n} \alpha_i [-\log P(h_i \mid b)]}{\sum_{i=1}^{n} -\log P(h_i \mid b)}$$

$$= t + \frac{\sum_{i=1}^{n} \alpha_i [-\log P(h_i \mid b)]}{\sum_{i=1}^{n} -\log P(h_i \mid b)}$$

$$\geq t$$

[from $\alpha_i \geq 0$ and $P(h_i \mid b) < 1$]

(ii) Suppose next $J(h_1, e \mid b), \ldots, J(h_n, e \mid b) < t$. Then, for any $i = 1, \ldots, n$, there is some $\beta_i > 0$ such that:

$$J(h_i, e \mid b) = \frac{\log P(h_i \mid e \wedge b) - \log P(h_i \mid b)}{-\log P(h_i \mid b)} = t - b_i$$

It follows that:

$$\log P(h_i \mid e) - \log P(h_i) = (t - b_i)[-\log P(h_i)]$$

By plugging this into (1) above, we obtain:

$$\begin{aligned} J(h_1 \wedge \ldots \wedge h_n, e \mid b) &= \frac{\sum_{i=1}^{n} (t - \beta_i)[-\log P(h_i \mid b)]}{\sum_{i=1}^{n} -\log P(h_i \mid b)} \\ &= \frac{t \sum_{i=1}^{n} -\log P(h_i \mid b) - \sum_{i=1}^{n} \beta_i [-\log P(h_i \mid b)]}{\sum_{i=1}^{n} -\log P(h_i \mid b)} \\ &= \frac{\sum_{i=1}^{n} \beta_i [-\log P(h_i \mid b)]}{\sum_{i=1}^{n} -\log P(h_i \mid b)} \\ &< t \end{aligned}$$

[from $\beta_i > 0$ and $P(h_i \mid b) < 1$] ■

4. Ordinal Equivalence

Suppose $W_1(h, e \mid b) = F_1(P(h \mid e \wedge b), P(h \mid b))$ and $W_2(h, e \mid b) = F_2(P(h \mid e \wedge b), P(h \mid b))$ are both continuous functions that are measures of epistemic worthiness. Then, they are ordinally equivalent to each other, i.e. for any two pairs $<h_i, e_i>$ and $<h_j, e_j>$, $W_1(h_i, e_i \mid b) <$ (or $=$, $>$) $W_1(h_j, e_j \mid b)$ if and only if $W_2(h_i, e_i \mid b) <$ (or $=$, $>$) $W_2(h_j, e_j \mid b)$, respectively.

Proof

Let $\log_{P(h_i|b)} P(h_j \mid b) = r$, so that $[P(h_i \mid b)]^r = P(h_j \mid b)$. It follows from $0 < P(h_i \mid b) < 1$ and $0 < P(h_j \mid b) < 1$ that $r > 0$. Since $J_1(h, e \mid b) = F_1(P(h \mid e \wedge b), P(h \mid b))$ and $J_2(h, e \mid b) = F_2(P(h \mid e \wedge b), P(h \mid b))$ are continuous functions, it suffices to show that the claim holds for any two pairs $<h_i, e_i>$ and $<h_j, e_j>$ such that $[P(h_i \mid b)]^q = P(h_j \mid b)$ where q is a positive rational number. Let $<m, n>$ be the smallest pair of positive integers such that $n/m = q$, so that $[P(h_i\ b)]^n$ – $[P(h_j \mid b)]^m$. Choose probabilistically independent (both on condition of b and on condition of $b \wedge e_i$) statements $h_1, \ldots, h_n$, and probabilistically independent (both on condition of b and on condition of $b \wedge e_j$) statements $h_{n+1}, \ldots, h_{n+m}$ such that:[1]

(i) $[P(h_i \mid b)]^n = [P(h_j \mid b)]^m$
(ii) $P(h_i \mid b) = P(h_1 \mid b) = \ldots = P(h_n \mid b)$
(iii) $P(h_j \mid b) = P(h_{n+1} \mid b) = \ldots = P(h_{n+m} \mid b)$
(iv) $P(h_i \mid e_i \wedge b) = P(h_1 \mid e_i \wedge b) = \ldots = P(h_n \mid e_i \wedge b)$
(v) $P(h_j \mid e_j \wedge b) = P(h_{n+1} \mid e_j \wedge b)) = \ldots = P(h_{n+m} \mid e_j \wedge b)$

It follows from (ii) and (iv) that $W_1(h_i, e_i \mid b) = W_1(h_1, e_i \mid b) = \ldots = W_1(h_n, e_i \mid b)$. So, by SCR:

$$W_1(h_i, e_i \mid b) = W_1(h_1 \wedge \ldots \wedge h_n, e_i \mid b) \tag{1}$$

Similarly, it follows from (iii), (v) and SCR that:

$$W_1(h_j, e_j \mid b) = W_1(h_{n+1} \wedge \ldots \wedge h_{n+m}, e_j \mid b) \tag{2}$$

Since $h_1, \ldots, h_n$ are probabilistically independent on condition of b, it follows from (ii) that $P(h_1 \wedge \ldots \wedge h_n \mid b) = [P(h_i \mid b)]^n$. Similarly, since $h_{n+1}, \ldots, h_{n+m}$ are probabilistically independent on condition of b, it follows from (iii) that $P(h_{n+1} \wedge \ldots \wedge h_{n+m} \mid b) = [P(h_j \mid b)]^m$. But $[P(h_i \mid b)]^n = [P(h_j \mid b)]^m$ from (i). So,

$$P(h_1 \wedge \ldots \wedge h_n \mid b) = P(h_{n+1} \wedge \ldots \wedge h_{n+m} \mid b) \tag{3}$$

Meanwhile, since $h_1, \ldots, h_n$ are probabilistically independent on condition of $e_i \wedge b$, it follows from (iv) that $P(h_1 \wedge \ldots \wedge h_n \mid e_i \wedge b) = [P(h_i \mid e_i \wedge b)]^n$. Similarly, since $h_{n+1}, \ldots, h_{n+m}$ are probabilistically independent on condition of $e_j \wedge b$, it follows from (v) that $P(h_{n+1} \wedge \ldots \wedge h_{n+m} \mid e_j \wedge b) = [P(h_j \mid e_j \mid b)]^m$. So,

$$P(h_1 \wedge \ldots \wedge h_n \mid e_i \wedge b) < (\text{or} =, >)\ P(h_{n+1} \wedge \ldots \wedge h_{n+m} \mid e_j \wedge b)$$

$$\text{iff } [P(h_i \mid e_i \wedge b)]^n < (\text{or} =, >)\ [P(h_j \mid e_j \mid b)]^m, \text{ respectively.} \tag{4}$$

Since $W_1(h, e \mid b) = F_1(P(h \mid e \wedge b), P(h \mid b))$ is an increasing function of $P(h \mid e \wedge b)$, it follows from (3) and (4) that:

$$W_1(h_1 \wedge \ldots \wedge h_n, e_i \mid b) < (\text{or} =, >)\ W_1(h_{n+1} \wedge \ldots \wedge h_{n+m}, e_j \mid b)$$

$$\text{iff } [P(h_i \mid e_i \wedge b)]^n < (\text{or} =, >)\ [P(h_j \mid e_j \wedge b)]^m, \text{ respectively.} \tag{5}$$

It follows from (1), (2) and (5) that:

$W_1(h_i, e_i \mid b) <$ (or $=, >$) $W_1(h_j, e_j \mid b)$

iff $[P(h_i \mid e_i \wedge b)]^n <$ (or $=, >$) $[P(h_j \mid e_j \wedge b)]^m$, respectively. (6)

By the same reasoning,

$W_2(h_i, e_i \mid b) <$ (or $=, >$) $W_2(h_j, e_j \mid b)$

iff $[P(h_i \mid e_i \wedge b)]^n <$ (or $=, >$) $[P(h_j \mid e_j \wedge b)]^m$, respectively. (7)

It follows from (6) and (7) that:

$W_1(h_i, e_i \mid b) <$ (or $=, >$) $W_1(h_j, e_j \mid b)$

iff $W_2(h_i, e_i \mid b) <$ (or $=, >$) $W_2(h_j, e_j \mid b)$, respectively ■.

Notes

1 For example, think of n urns of colored marbles, for each of which the probability of drawing a red marble is the same as $P(h_i \mid b)$, and m urns of colored marbles, for each of which the probability of drawing a red marble is the same as $P(h_j \mid b)$. To satisfy the conditions $P(h_i \mid e_i \wedge b) = P(h_i \mid b)$ and $P(h_j \mid e_j \wedge b) = P(h_j \mid b)$ of the theorem (in addition to (i) through (v)), the n urns must have nothing to do with e_i and the m urns must have nothing to do with e_j.

2 For example, think of n urns of colored marbles, for each of which the probability of drawing a red marble is the same as $P(h_i \mid b)$, but given the evidence e_i that the n urns belong to a certain type, the probability of drawing a red marble is the same as $P(h_i \mid e_i \wedge b)$. Similarly, think of m urns of colored marbles, for each of which the probability of drawing a red marble is the same as $P(h_j \mid b)$, but given the evidence e_j that the m urns belong to a certain other type, the probability of drawing a red marble is the same as $P(h_j \mid e_j \wedge b)$

5 A Bayesian Case for Skepticism

5.1 No Closure under Logical Entailment

This chapter investigates the implications of the dual-component format of epistemic evaluation for Cartesian skepticism. Some are good news for the defenders of the natural world hypothesis, but in the end the dual-component format supports Cartesian skepticism. The main point of the chapter is to challenge Bayesian complaisance by pointing out that we cannot take refuge in subjective prior probabilities, i.e. we cannot hold on to the natural world hypothesis even if we assign a high subjective prior probability to the natural world hypothesis.

I start with some points that are in favor of the natural world hypothesis. First, unlike the standard Bayesian measures, the measure $J(h, e \mid b)$ of epistemic worthiness does not automatically rank a disjunction above its disjunct. Recall that the standard Bayesian measure $P(h \mid e \wedge b)$ is a measure of how safe (or how risky, to put it negatively) it is to accept h. Suppose h_1 and h_2 are rival hypotheses that are mutually exclusive, and neither of them is refuted with absolute certainty. To put it formally, $P(h_1 \wedge h_2 \mid e \wedge b) = 0$ while $P(h_1 \mid e \wedge b) > 0$ and $P(h_2 \mid e \wedge b) > 0$. It follows immediately that $P(h_1 \mid e \wedge b) < P(h_1 \vee h_2 \mid e \wedge b)$ and $P(h_2 \mid e \wedge b) < P(h_1 \vee h_2 \mid e \wedge b)$. In other words, accepting the disjunction $h_1 \vee h_2$ is safer than accepting h_1 by itself or accepting h_2 by itself. This is in support of Cartesian skepticism that urges us to remain non-committal between the natural world hypothesis h_{NW} and the BIV hypothesis h_{BIV} because h_{NW} and h_{BIV} are mutually exclusive and neither of them is refuted with absolute certainty. It is safer to remain non-committal than to accept h_{NW} by itself, or $P(h_{NW} \mid e \wedge b) < P(h_{NW} \vee h_{BIV} \mid e \wedge b)$ to put it formally.[1]

The measure $J(h, e \mid b)$ of epistemic worthiness, in contrast, does not automatically rank a disjunction above its disjunct even if the disjuncts are mutually exclusive and neither is refuted with absolute certainty. The reason is the additional component in the format. The dual-component format takes into account not just the degree of safety of accepting h, but also the amount of information that h carries. This makes the measure $J(h, e \mid b)$ an increasing function of $P(h \mid e \wedge b)$ and a decreasing function of $P(h \mid b)$. As a result, $J(h_{NW}, e \mid b)$ may be greater than $J(h_{NW} \vee h_{BIV}, e \mid b)$ despite $P(h_{NW} \mid e \wedge b) < P(h_{NW} \vee h_{BIV} \mid e \wedge b)$, if $P(h_{NW} \mid b)$ is sufficiently smaller than $P(h_{NW} \vee h_{BIV} \mid b)$. There is therefore a hope that the dual-component format may endorse the natural world hypothesis despite the greater safety of remaining non-committal between the two hypotheses.

The possibility that the measure $J(h, e \mid b)$ may rank a disjunct above a disjunction is related to another point in favor of the natural world hypothesis, viz. epistemic justification by the dual-component format is not closed under logical entailment.[2] It is possible, in other words, that $J(h_1, e \mid b) \geq t$ and h_1 logically entails h_2, but $J(h_2 \mid e \wedge b) < t$. The reason is essentially the same as stated above. If h_1 logically entails h_2, then $P(h_1 \mid e \wedge b) \leq P(h_2 \mid e \wedge b)$. It is therefore never safer to accept the hypothesis h_1 than to accept its logical consequence h_2. As a result, epistemic justification in the Lockean format is closed under logical entailment in that if $P(h_1 \mid e \wedge b) \geq t$ and h_1 logically entails h_2, then $P(h_2 \mid e \wedge b) \geq t$.[3] As discussed in the previous chapter, epistemic justification in the Lockean format is not closed under *multiple-premise* logical entailment, but it is closed under *single-premise* logical entailment.

In contrast, epistemic justification in the dual-component format is not closed under single-premise logical entailment because the measure $J(h, e \mid b)$ of epistemic worthiness takes into account not just the degree of safety of accepting h. It also takes into account the amount of information that h carries, which is a decreasing function of $P(h \mid b)$. So, $J(h_1, e \mid b)$ can be greater than $J(h_2 \mid e \wedge b)$ even if h_1 logically entails h_2, if $P(h_1 \mid b)$ is sufficiently smaller than $P(h_2 \mid b)$. It is therefore possible that $J(h_1, e \mid b) \geq t$ and h_1 logically entails h_2 but $J(h_2, e \mid b) < t$ for some threshold t.

It is significant that single-premise epistemic closure (hereafter simply "epistemic closure") does not hold in the dual-component format because epistemic closure has been used against an effort to contain Cartesian skepticism. Take the particular statement h_H that here is a hand. If I have the sensory evidence e of apparently seeing a hand in front of me and I notice nothing suspicious about the situation, then it seems fine for me to accept that here is a hand. This seems different from the abstract question of whether I am a brain in a vat. Guided by this idea, some epistemologists explored the possibility of defending particular claims about the natural world such as h_H, while conceding that we do not have justification for the denial of the BIV hypothesis $\neg h_{BIV}$.[4] If successful, it will contain Cartesian skepticism in a small area of metaphysical hypotheses. The obstacle to the containment strategy, however, is that h_H logically entails $\neg h_{BIV}$, i.e. if it is true that here is a hand, then it follows logically that the subject is not a brain in a vat.[5] So, by closure under logical entailment, if there is justification for accepting h_H, then there is also justification for denying h_{BIV}. Equivalently, if there is no justification for denying h_{BIV}, then there is no justification for accepting h_H. Epistemic closure undercuts the effort to contain Cartesian skepticism.

It is tempting to reject epistemic closure, but that is often considered a last resort—even a desperate move.[6] When some people do abandon epistemic closure, it is usually with regard to *knowledge*, and not epistemic justification. Some prominent epistemologists (Dretske 1971; Nozick 1981) have denied closure of knowledge under logical entailment, but their arguments depend on their particular analyses of knowledge, such as the relevant-alternative analysis and the tracking analysis, and are difficult to translate into arguments against closure of epistemic justification.[7] One reason for the reluctance to abandon the closure of epistemic justification is the Lockean format of epistemic evaluation. There is no way to get around epistemic closure in the Lockean format. Replacing the Lockean format by the dual-component format makes it possible to contain Cartesian skepticism because justification in the dual-component format is not closed under logical entailment. There is, however, a serious concern among epistemologists that abandoning epistemic closure deprives us of the ability to expand the body of beliefs by deductive logic. For example, even if there is justification for adding the statement *that all emeralds are green* to our body of beliefs, we may not have justification for adding the statement *that the next emerald we observe is green*. It seems to be a serious blow to our epistemic practice to lose the ability to expand the body of beliefs by deductive logic.

It turns out the fear of losing deductive expansion is groundless. We can expand the body of beliefs by deductive logic even if epistemic justification is not closed under logical entailment (Moretti and Shogenji 2017). Suppose $J(h_1, e \mid b) \geq t$ and h_1 logically entails h_2, but $J(h_2, e \mid b) < t$. In other words, there is justification for adding h_1 to b, but not its logical consequence h_2. This is actually no obstacle to deductive expansion because once we add h_1 to b to make the body of beliefs $b \wedge h_1$, we can expand it further to make it $b \wedge h_1 \wedge h_2$. Note that the bodies of beliefs $b \wedge h_1$ and $b \wedge h_1 \wedge h_2$ are logically equivalent when h_1 logically entails h_2. The addition of h_2 to $b \wedge h_1$ is an *analytic expansion* that is free of risk and devoid of additional information.[8]

So, after adding h_1 to b first, we can add h_2 to $b \wedge h_1$. Note also that we can achieve the same result in one step by the principle of epistemic closure under *logical equivalence*, which is much weaker than the principle of epistemic closure under logical entailment.

> **Epistemic Closure under Logical Entailment:** If $J(h_1, e \mid b) \geq t$ and h_1 logically entails h_2, then $J(h_2, e \mid b) \geq t$.

> **Epistemic Closure under Logical Equivalence:** If $J(h_1, e \mid b) \geq t$, and h_1 and h_2 are logically equivalent, then $J(h_2, e \mid b) \geq t$.[9]

Epistemic closure under logical equivalence is trivial and should hold under any format of epistemic evaluation,[10] but it is sufficient for expanding the body of beliefs by logical deduction. Suppose h_1 logically entails h_2. It follows immediately that h_1 and $h_1 \wedge h_2$ are logically equivalent. So, if there is justification for adding h_1 to the body of beliefs b, then there is also justification for adding $h_1 \wedge h_2$ to b by epistemic closure under logical equivalence.

There may be some concern that this allows an epistemically unworthy statement to piggyback on an epistemically worthy statement that logically entails it, but that is harmless provided the subject consults the whole body of beliefs in her future deliberations, for example, consulting the statement that all emeralds are green, instead of selectively consulting the statement that the next emerald we observe is green. This is also true of the hand hypothesis h_H and the denial of the BIV hypothesis $\neg h_{BIV}$ that is logically entailed by it. If there is justification for accepting h_H, then there is justification for accepting $h_H \wedge \neg h_{BIV}$ that is logically equivalent to it. We can therefore add $\neg h_{BIV}$ to the body of beliefs together with h_H, but it is harmless provided the subject consults the whole body of beliefs in her future deliberations, avoiding selectively consulting $\neg h_{BIV}$ in disregard of h_H.

5.2 Consistency with Epistemic Practice

So, the crucial issue in epistemic closure under logical entailment is not deductive expansion of the body of beliefs. Deductive expansion is secured by the weaker principles of epistemic closure under logical equivalence (or by analytic expansion). Epistemic closure under logical entailment has consequences that are stronger than necessary for deductive expansion. Deductive expansion is free of risk and devoid of additional information, but expansion by epistemic closure under logical entailment carries both risk and gain. This section shows that denying it as a general principle is actually consistent with our pre-theoretical judgment, viz. in some cases we are reluctant to add h_2 to b even if there is justification for adding h_1 to b and h_1 logically entails h_2. Of course, pre-theoretical judgment is not reason for accepting or rejecting a theory in meliorative epistemology. My point is that there is no need to revise our epistemic practice in light of the dual-component format of epistemic evaluation with regard to epistemic closure.

I want to illustrate this with a variant of the widely discussed case, *the feminist bank teller Linda*, from the psychology of probabilistic fallacies (Tversky and Kahneman 1983). The following are the setting of the case. The participants in the experiment receive the information e about Linda, and are asked to evaluate the two statements h_1 (a conjunction) and h_2 (one of the conjuncts in h_1):

e: Linda is 31 years old, single, outspoken and very bright. She majored in philosophy. As a student, she was deeply concerned with issues of discrimination and social justice, and also participated in anti-nuclear demonstrations.

h_1: Linda is a bank teller and is active in the feminist movement.

h_2: Linda is a bank teller.

The original Linda question asks which of the two statements, h_1 and h_2, is *more probable* given the information e, and to the dismay of logicians many people answer that h_1 is more probable than h_2 thereby committing the *conjunction fallacy*, i.e. the fallacy of assigning a higher probability to the conjunction than to one of its conjuncts. Since h_1 logically entails h_2, $P(h_1 \mid e \wedge b)$ cannot be greater than $P(h_2 \mid e \wedge b)$.

Consider now the following variant of the Linda question: Which of the two statements, h_1 and h_2, is *more worthy* of inclusion in the body of beliefs given the information e? Less formally, which of the two statements are you *more willing* to accept given the information e? Many participants' answer to the original question in favor of h_1 indicates their greater willingness to accept *that Linda is a feminist bank teller* than to accept *that Linda is a bank teller* given the information e. If you are like most participants in the experiment and if you set aside theoretical considerations, you will also be more willing to accept h_1 than you will h_2. The point is that the denial of epistemic closure under logical entailment is more consistent with our pre-theoretical judgment. Though meliorative epistemology does not aim to accommodate pre-theoretical judgments, consistency with them is still noteworthy—there is no need to revise our epistemic practice with regard to epistemic closure.

To reinforce the point, I want to show formally that given the information e and relative to the body of beliefs b typical among people living in the modern world, h_1 is epistemically more worthy of inclusion in b than h_2 is, or $J(h_1, e \mid b) > J(h_2, e \mid b)$. The general reason is the same as before: The greater potential gain in truth that h_1 delivers outweighs the higher risk taken when h_1 is added to b.[11] More specifically, the information e raises the probability *that Linda is a feminist bank teller*, or $P(h_1 \mid e \wedge b) > P(h_1 \mid b)$, while e lowers the probability *that Linda is a bank teller*, or $P(h_2 \mid e \wedge b) < P(h_2 \mid b)$. As a result, $J(h_1, e \mid b)$ is positive while $J(h_2, e \mid b)$ is negative:

$$J(h_1, e \mid b) = \frac{\log P(h_1 \mid e \wedge b) - \log P(h_1 \mid b)}{-\log P(h_1 \mid b)} > 0.$$

$$J(h_2, e \mid b) = \frac{\log P(h_2 \mid e \wedge b) - \log P(h_2 \mid b)}{-\log P(h_2 \mid b)} < 0.$$

It follows that $J(h_1, e \mid b) > J(h_2, e \mid b)$. In other words, given e, the conjunction h_1 is epistemically more worthy of inclusion in b than the conjunct h_2 is.[12]

So, we have good reason in meliorative epistemology to reject epistemic closure under logical entailment, and the rejection is consistent with our pre-theoretical judgment in the Linda case. However—and this is the key point—once we add h_1 to the body of beliefs b, we can also add h_2 to the expanded body of beliefs $b \wedge h_1$ because h_1 logically entails h_2. This may look suspiciously similar to the inverse backdoor endorsement discussed in the previous chapter, viz. there is no justification for adding h_2 to b given e, but we can add it to b anyways through the backdoor by first adding the conjunction h_1 that logically entails h_2. However, in the Linda case we are not combining two conjuncts that are both epistemically unworthy and irrelevant to each other. The other conjunct—that Linda is active in the feminist movement—is of high epistemic worthiness given e. Otherwise, i.e. if both conjuncts are epistemically unworthy and irrelevant to each other, then the Inverse Backdoor Constraint blocks the addition of the conjunction.

Of course, willingness to accept some conjunction more than its conjunct is just one aspect of our epistemic practice, and the question remains whether abandoning epistemic closure under logical entailment is consistent with some other aspects of our epistemic practice. Again, the reason for asking this question is not the desire to make the theory consistent with our epistemic practice, but to see whether some aspects of our epistemic practice is in need of revision. For example, even if there is justification for adding the general statement *that all emeralds are green* to our body of beliefs, there may not be justification for adding its instantiation *that the next emerald we observe is green*. This may seem inconsistent with our epistemic practice, but it is only inconsistent with the common *interpretation* of our epistemic practice. It is unusual in this type of situation to add the particular statement without also adding the general statement. As long as the general statement is also added to the body of beliefs, our epistemic practice is well supported by analytic expansion and epistemic closure under logical equivalence.

Another point of concern is the reasoning by modus tollens. We often deny or question a statement because it logically entails some statement we deny or question. With regard to denial, the point about analytic expansion and epistemic closure under logical equivalence also applies. Suppose h_1 logically entails h_2 and we have good reason to deny h_2, i.e. there is justification for adding $\neg h_2$ to b. In the absence of epistemic closure under logical entailment, it does not follow from these that there is justification for adding $\neg h_1$ to b. However, once we add $\neg h_2$ to b, we can also add $\neg h_1$ to $b \wedge \neg h_2$ by analytic expansion because if h_1 logically entails h_2, then $\neg h_2$ logically entails $\neg h_1$. We can also add the conjunction $\neg h_1 \wedge \neg h_2$ directly, which is logically equivalent to the conjunct $\neg h_2$. This is consistent with our epistemic practice because it is uncommon to add $\neg h_1$ by modus tollens without also adding $\neg h_2$ that prompted the denial of h_1.

However, the same point does not apply to modus tollens from the questioning of a logical consequence. Suppose h_1 logically entails h_2 and we question h_2. In other words, there is no justification for adding h_2 to b though there is no justification for adding its negation $\neg h_2$ to b either. If epistemic justification is closed under logical entailment, it follows by modus tollens that there is no justification for adding h_1 to b. This is the form of reasoning that undercuts the effort to contain Cartesian skepticism, where h_1 is a particular hypothesis about the natural world—e.g. that here is a hand—while h_2 is the denial of the BIV hypothesis. Even in the presence of the sensory experience e of apparently seeing a hand in front of us, the skeptic exploits our doubt about h_2 to question the epistemic status of h_1 by modus tollens.

Many epistemologists are swayed by the reasoning, but they should not be. Once we abandon epistemic closure under logical entailment, as we should, even if there is no justification for adding h_2 to b and h_1 logically entails h_2, there may still be justification for adding h_1 to b. The suggestion is to abandon an inference pattern that is widely accepted among epistemologists, but abandoning it is not inconsistent with our actual epistemic practice. Indeed, on the intuitive level most people find it unproblematic to accept that here is a hand given the relevant sensory experience, though they may concede their inability to justify the denial of the BIV hypothesis. Of course, whether we actually have justification for accepting particular statements about the natural world is a substantive question to be addressed in the subsequent sections and chapters. The point at this juncture is that the absence of justification for denying the BIV hypothesis does not force us to conclude that there is no justification for accepting particular statements about the natural world.

5.3 Supersedure

I pointed out two implications of the dual-component format of epistemic evaluation so far that are in favor of the natural world hypothesis. The measure $J(h, e \mid b)$ of epistemic worthiness does not automatically rank a disjunction above its disjunct. So, it does not automatically rank the disjunction of the natural world hypothesis and the BIV hypothesis above the natural world hypothesis. Also, epistemic justification is not closed under logical entailment. So, there is no need to argue against the BIV hypothesis for the purpose of defending particular hypotheses about the natural world, e.g. the hypothesis h_H that here is a hand.

In fact it looks like there is epistemic justification for accepting h_H if the epistemic subject has the relevant reflective visual evidence e, i.e. if it appears to her visually that here is a hand. This becomes clear in the following rewriting of $J(h, e \mid b)$:

$$J(h,e \mid b) = \frac{\log P(h \mid e \wedge b) - \log P(h \mid b)}{-\log P(h \mid b)}$$
$$= 1 - \frac{\log P(h \mid e \wedge b)}{\log P(h \mid b)}.$$

Since the body of beliefs b is empty in the context of Cartesian skepticism, the two determinants of epistemic worthiness of h_H are $P(h_H \mid e)$ and $P(h_H)$. Of these two, $P(h_H)$ is very low since h_H is a very particular statement. This makes $\log P(h_H)$ a negative number of large magnitude. Meanwhile, $P(h_H \mid e) = [P(e \mid h_H) \times P(h_H)]/P(e)$ is much higher than $P(h_H)$ because $P(e \mid h_H)$ is much higher than $P(e)$.[13] This makes $\log P(h_H \mid e)$ a negative number of small magnitude. As a result, the ratio $\log P(h_H \mid e)/\log P(h_H)$ is a positive number much smaller than one, and thus $J(h_H, e \mid b)$ is not far away from the maximum value 1. We can therefore make a strong case that there is epistemic justification for accepting h_H, or $J(h_H, e \mid b) \geq t$, unless we set the threshold of sufficiency t extremely high.

The main task of this chapter, however, is to show that there is actually no justification in the dual-component format of epistemic evaluation for accepting particular statements about the natural world, such as h_H, even in the presence of the relevant evidence, and even if the threshold of sufficiency t is not extremely high. To be a little more precise, there is prima facie justification, but not all-things-considered justification, for accepting those statements because they are *superseded*, i.e. there are statements that are epistemically more worthy than they are, and once the more worthy statements are accepted, there is no longer justification for accepting them. Before explaining the concept of supersedure in more detail, I want to briefly go over the concept of defeat that resembles supersedure in certain ways.

It is widely acknowledged that epistemic justification that falls short of absolute certainty is defeasible by new empirical evidence. This is sensible and consistent with both the Lockean format and the dual-component format. To begin with the Lockean format, even if the initial body of evidence e together with the body of beliefs b makes the hypothesis h highly probable and thus very safe to accept, some additional evidence may reveal that h is not so probable. To express it formally, since the conditional probability is not monotonic with regard to the expansion of the body of evidence, it is possible that $P(h \mid e \wedge b) > P(h \mid e \wedge e' \wedge b)$, where e is the original body of evidence and e' is the additional piece of evidence. It is therefore possible that $P(h \mid e \wedge b) \geq t$ while $P(h \mid e \wedge e' \wedge b) < t$ for some threshold t. This means that epistemic justification in the Lockean format is defeasible.[14]

Epistemic justification by the dual-component format is also defeasible by new empirical evidence. Even if the initial body of evidence e together with the body of beliefs makes the hypothesis h highly probable and its high degree of safety relative to the amount of information it carries makes h epistemically worthy, the additional evidence may reveal that h is not so probable and its risk relative to the amount of information it carries may make h epistemically unworthy. We can see this formally as follows. First, as just noted, it is possible that $P(h \mid e \wedge b) > P(h \mid e \wedge e' \wedge b)$ because the conditional probability is not monotonic with regard to the expansion of the body of evidence. Meanwhile, $J(h, e \mid b) > J(h, e \wedge e' \mid b)$ when $P(h \mid e \wedge b) > P(h \mid e \wedge e' \wedge b)$, because the other determinant $P(h \mid b)$ is not affected by the expansion of the body of evidence. It is therefore possible that $J(h, e \mid b) \geq t$ while $J(h, e \wedge e' \mid b) < t$ for some threshold t.

Though it is not as widely acknowledged as empirical defeat, there is also the possibility of a priori defeat. It is arguable at least that epistemic justification is defeasible by new theoretical development. For example, the introduction of a novel hypothesis can change the probability distribution over the previously known hypotheses from P to P^*.[15] It is therefore possible that

$P(h \mid e \wedge b) > P^*(h \mid e \wedge b)$ for some previous known hypothesis h, and thus $P(h \mid e \wedge b) \geq t$ while $P^*(h \mid e \wedge b) < t$ for some threshold t. If this is possible, epistemic justification in the Lockean format is defeasible by the introduction of a novel hypothesis. We may call it a priori defeat since no new empirical evidence is needed for the defeat. Epistemic justification in the dual-component format is also susceptible to a priori defeat because it is possible that $J(h, e \mid b) > J^*(h, e \mid b)$, where $J(h, e \mid b)$ is the degree of epistemic worthiness based on the previous probability distribution P and $J^*(h, e \mid b)$ is the degree of epistemic worthiness based on the revised probability distribution P^*. It is therefore possible that $J(h, e) \geq t$ while $J^*(h, e \mid b) < t$ for some threshold t. There may be some objection to the concept of a priori defeat based on a strong version of the omniscience assumption, viz. that all possible hypotheses are known to the evaluator. It is not my objective, however, to defend the possibility of a priori defeat here. I mention it only to distinguish it from the concept of supersedure, which is the main focus of the section.

So, what is supersedure? As in empirical defeat and a priori defeat, the apparent justification for h is revoked, but supersedure is not brought about by a new piece of evidence or new theoretical development. The hypothesis h turns out to be not justified all things considered because there is a previously known hypothesis h^* that is epistemically more worthy, and once h^* is added to the body of beliefs b, there is no justification for adding h to the expanded body of beliefs $b \wedge h^*$. Here is a formal statement of supersedure in the dual-component format of epistemic evaluation:

> **Supersedure in the Dual-Component Format:** Suppose $J(h, e \mid b) \geq t$. The hypothesis h is superseded by the hypothesis h^* just in case $J(h, e \mid b) < J(h^*, e \mid b)$ and $J(h, e \mid b \wedge h^*) < t$.

Supersedure is not unique to the dual-component format. It also occurs in the Lockean format of epistemic evaluation:

> **Supersedure in the Lockean Format:** Suppose $P(h \mid e \wedge b) \geq t$. The hypothesis h is superseded by the hypothesis h^* just in case $P(h \mid e \wedge b) < P(h^* \mid e \wedge b)$ and $P(h \mid e \wedge b \wedge h^*) < t$.

I want to explain supersedure in the Lockean format, first, because the Lockean format is more familiar to Bayesian epistemologists.

Consider a six-faced die, and assume that the body of evidence e together with the body of beliefs b has already established that the die is biased in the following way: $P(x = 1 \mid e \wedge b) = 1/15$, $P(x = 2 \mid e \wedge b) = 2/15$, and $P(x = 3 \mid e \wedge b) = P(x = 4 \mid e \wedge b) = P(x = 5 \mid e \wedge b) = P(x = 6 \mid e \wedge b) = 3/15$, where x is the number that appears on the top face of the die. If the threshold of sufficiency t is 13/15, then there is justification for accepting $\neg(x = 2)$ for the next roll because $P(\neg(x = 2) \mid e \wedge b) = 13/15 \geq t$. However, the hypothesis $\neg(x = 2)$ is superseded by the hypothesis $\neg(x = 1)$ because

$P(\neg(x = 1) \mid e \wedge b) = 14/15$ is greater than $P(\neg(x = 2) \mid e \wedge b) = 13/15$, and once $\neg(x = 1)$ is added to b, $P(\neg(x = 2) \mid e \wedge b \wedge \neg(x = 1)) = 12/14$ falls short of the threshold of sufficiency t. So, there is prima facie justification for accepting the hypothesis $x = 2$, but not all-things-considered justification.

The situation is somewhat similar to the lottery case discussed in Chapter 4, where a hypothesis that is initially highly probable becomes much less probable once other statements that are equally probable are added to the body of beliefs b. The difference this time is that there is good reason in the Lockean format for adding the hypothesis $\neg(x = 1)$ *before* adding the hypothesis $\neg(x = 2)$ because $P(\neg(x = 1) \mid e \wedge b)$ is greater than $P(\neg(x = 2) \mid e \wedge b)$. So, despite its prima facie justification, the hypothesis $\neg(x = 2)$ is superseded by the hypothesis $\neg(x = 1)$, and this is not due to any new evidence or any new theoretical development.

Supersedure in the dual-component format occurs in a similar way. Suppose two hypotheses h_1 and h_2 have the same low probability $P(h_1 \mid b) = P(h_2 \mid b) \approx 0$ on condition of the body of beliefs b alone. If the body of evidence e makes both of them highly probable but is slightly in favor of h_2, then it is possible that $J(h_2, e \mid b) > J(h_1, e \mid b) \geq t$ for some threshold t. In other words, there is justification for accepting each of the two hypotheses, but we have good reason to accept h_2 before accepting h_1. Suppose further that the addition of h_2 to b reduces the epistemic worthiness of h_1 because either $P(h_1 \mid h_2 \wedge b) > P(h_1 \mid b)$, which reduces the amount of information h_1 carries, or $P(h_1 \mid e \wedge h_2 \wedge b) < P(h_1 \mid e \wedge b)$, which reduces the degree of safety, to the extent that $J(h_1, e \mid b \wedge h_2) < t$. The hypothesis h_1 is then superseded by the hypothesis h_2.

Supersedure in the dual-component format sometimes occurs in a stark way, where the acceptance of the epistemically more worthy hypothesis h_2 does not just reduce the epistemic worthiness of the hypothesis h_1 below the threshold of sufficiency t, but eliminates h_1 entirely because h_1 is incompatible with h_2. This may seem odd at first because supersedure requires that there is justification for accepting each of the hypotheses h_1 and h_2. How can that happen if h_1 and h_2 are incompatible? It turns out this is possible for any threshold of sufficiency t less than one (Wójtowicz and Bigaj 2016). It is not difficult to see the reason from the following formulation of $J(h, e \mid b)$.

$$J(h_1, e \mid b) = 1 - \frac{\log P(h_1 \mid e \wedge b)}{\log P(h_1 \mid b)}.$$

$$J(h_2, e \mid b) = 1 - \frac{\log P(h_2 \mid e \wedge b)}{\log P(h_2 \mid b)}.$$

First, $P(h_1 \mid e \wedge b) + P(h_2 \mid e \wedge b)$ cannot exceed one when h_1 and h_2 are incompatible, but each of them can still be close to 1/2. So, if we use the log base 2, both $\log_2 P(h_1 \mid e \wedge b)$ and $\log_2 P(h_2 \mid e \wedge b)$ may be only slightly less than -1. Meanwhile, as $P(h_1 \mid b)$ and $P(h_2 \mid b)$ both approach zero, $\log_2$

$P(h_1 \mid b)$ and $\log_2 P(h_2 \mid b)$ decreases with no finite lower bound. As a result, the two ratios $\log_2 P(h_1 \mid e \wedge b)/\log_2 P(h_1 \mid b)$ and $\log_2 P(h_2 \mid e \wedge b)/\log_2 P(h_2 \mid b)$ both approach zero, so that $J(h_1, e \mid b)$ and $J(h_2, e \mid b)$ approach one to meet the condition $J(h_1, e \mid b) \geq t$ and $J(h_2, e \mid b) \geq t$ for any threshold of sufficiency t less than one. To put this intuitively, large amounts of information they carry can counterbalance their modest risks to raise their epistemic worthiness as high as needed to meet any threshold of sufficiency less than one.

There is no need to worry that we may end up with incompatible statements in the body of beliefs. Once one of the incompatible statements is added to the body of beliefs, there is no justification for adding the other. The same point also applies to the Lockean format of epistemic evaluation. It is customary to set the threshold of sufficiency t above 1/2 to prevent that both h and its negation $\neg h$ reach t. However, if that is the only reason for setting t above 1/2, it is unnecessary because once one of them, say h, is accepted, the probability of its negation $P(\neg h \mid e \wedge b \wedge h)$ is reduced to zero to prevent the addition of $\neg h$.[16] Besides, setting t above 1/2 is not enough for preventing each of the jointly inconsistent statements $h_1, \ldots, h_n$ from reaching the threshold, as we saw in the lottery case in Chapter 4. We cannot avoid such cases by simply raising t higher because for any threshold t less than one, we can increase n to meet it. The proper way to handle the lottery challenge in the Lockean format is to point out that once sufficiently many of the statements are added to the body of beliefs, there is no more justification for adding any of the remaining statements.[17]

5.4 A Case for Cartesian Skepticism

With the concept of supersedure in hand, I now return to Cartesian skepticism. As noted earlier, we can make a strong case for accepting particular statements about the natural world such as h_H (that here is a hand) given the appropriate evidence e, in the sense that $J(h_H, e \mid b) \geq t$. However, $J(h_H, e \mid b) \geq t$ only establishes prima facie justification, and may not be enough for all-things-considered justification, because h_H may be superseded by some statement that is epistemically more worthy. The most obvious concern is the virtual hand hypothesis h_{VH}. To spell it out, the hypothesis h_{VH} states that I am a brain in a vat connected to a supercomputer and the supercomputer is feeding me the sensory experience e of apparently seeing a hand in front of me. The two hypotheses h_H and h_{VH} are incompatible, but as explained in the previous section, there may be justification for accepting each of them, i.e. $J(h_H, e \mid b) \geq t$ and $J(h_{VH}, e \mid b) \geq t$. Indeed, we can make a strong case for the latter just as we can for the former. Since the body of beliefs b is empty in the context of Cartesian skepticism, the two determinants of the epistemic worthiness of h_{VH} are $P(h_{VH} \mid e)$ and $P(h_{VH})$. Of these two, $P(h_{VH})$ is very low since h_{VH} is a very particular statement. This makes the amount of information that h_{VH} carries very large. Meanwhile, $P(h_{VH} \mid e)$ is much higher than $P(h_{VH})$, and this makes the risk of accepting h_{VH} small relative to the amount of information that h_{VH} carries.

It is therefore arguable that $J(h_{VH}, e \mid b) \geq t$ for some sensible threshold t. If there is justification for each of the two statements, as it appears, then one with a higher degree of epistemic worthiness supersedes the other. So, which is greater, $J(h_H, e \mid b)$ or $J(h_{VH}, e \mid b)$?

It turns out the answer solely depends on their prior probabilities. To see why, we rewrite the measure $J(h, e \mid b)$ of epistemic worthiness as follows:

$$
\begin{aligned}
J(h,e \mid b) &= \frac{\log P(h \mid e \wedge b) - \log P(h \mid b)}{-\log P(h \mid b)} \\
&= \frac{\log\left[P(h \wedge e \mid b) / P(e \mid b)\right] - \log P(h \mid b)}{-\log P(h \mid b)} \\
&= \frac{\left[\log P(h \wedge e \mid b) - \log P(e \mid b)\right] - \log P(h \mid b)}{-\log P(h \mid b)} \\
&= \frac{\left[\log P(h \wedge e \mid b) - \log P(h \mid b)\right] - \log P(e \mid b)}{-\log P(h \mid b)} \\
&= \frac{\log\left[P(h \wedge e \mid b) / P(h \mid b)\right] - \log P(e \mid b)}{-\log P(h \mid b)} \\
&= \frac{\log P(e \mid h \wedge b) - \log P(e \mid b)}{-\log P(h \mid b)}
\end{aligned}
$$

Since the body of beliefs b is empty in the context of Cartesian skepticism, the three determinants of the epistemic worthiness of h in this formulation of $J(h, e \mid b)$ are the two prior probabilities $P(h)$ and $P(e)$, and the likelihood $P(e \mid h)$. So, we are comparing the following two formulas:

$$
J(h_{\mathrm{H}}, e \mid b) = \frac{\log P(e \mid h_{\mathrm{H}}) - \log P(e)}{-\log P(h_{\mathrm{H}})}
$$

$$
J(h_{\mathrm{VH}}, e \mid b) = \frac{\log P(e \mid h_{\mathrm{VH}}) - \log P(e)}{-\log P(h_{\mathrm{VH}})}
$$

The key in the comparison is that the two likelihoods $P(e \mid h_H)$ and $P(e \mid h_{VH})$ are identical since the supercomputer is programed to generate the same sensory experience in the brain in a vat that the brain in a skull receives in the natural world. Since $P(e)$ is the same between the two formulas, the relative worthiness of the two hypotheses solely depends on their prior probabilities $P(h_H)$ and $P(h_{VH})$. Since the numerators are positive and identical, their respective epistemic worthiness is an inverse function of the denominator, $-\log P(h_H)$ and $-\log P(h_{VH})$. So, their respective degree of epistemic worthiness is a direct function of their prior probability, $P(h_H)$ and $P(h_{VH})$.[18] In short, $J(h_H, e \mid b) < J(h_{VH}, e \mid b)$ and thus h_H is superseded by h_{VH} if and only if $P(h_H) < P(h_{VH})$.

It looks like the dispute over supersedure boils down to the prior probabilities $P(h_H)$ and $P(h_{VH})$. Recall, however, that the skeptic is not trying to show that the BIV hypothesis is preferable to the natural world hypothesis. The skeptic is only urging us to withhold judgment. This is also true of the more specific hypotheses h_{H} and h_{VH}. The hypothesis that supersedes the hand hypothesis h_{H}, according to the skeptic, is not the virtual hand hypothesis h_{VH}, but the disjunction $h_{\mathrm{H}} \vee h_{\mathrm{VH}}$ of the two hypotheses. So, we need to compare $J(h_{\mathrm{H}}, e \mid b)$ and $J(h_{\mathrm{H}} \vee h_{\mathrm{VH}}, e \mid b)$.

The reasoning proceeds in the same way as above. Since the body of beliefs b is empty in the context of Cartesian skepticism, we are comparing the following two formulas:

$$J(h_{\mathrm{H}}, e \mid b) = \frac{\log P(e \mid h_{\mathrm{H}}) - \log P(e)}{-\log P(h_{\mathrm{H}})}$$

$$J(h_{\mathrm{H}} \vee h_{\mathrm{VH}}, e \mid b) = \frac{\log P(e \mid h_{\mathrm{H}} \vee h_{\mathrm{VH}}) - \log P(e)}{-\log P(h_{\mathrm{H}} \vee h_{\mathrm{VH}})}$$

There are three determinants of epistemic worthiness to compare. To begin with the easiest, the prior probability of the evidence $P(e)$ is the same between the two formulas. Next, the prior probability of the hand hypothesis $P(h_{\mathrm{H}})$ is smaller than that of the disjunction $P(h_{\mathrm{H}} \vee h_{\mathrm{VH}})$ because h_{H} and h_{VH} are incompatible and h_{VH} is not refuted with absolute certainty. The remaining question is the third determinant, their likelihoods $P(e \mid h_{\mathrm{H}})$ and $P(e \mid h_{\mathrm{H}} \vee h_{\mathrm{VH}})$.

It turns out they are identical. This is not surprising on some reflection. Whether h_{H} is true or h_{VH} is true makes no difference in the conditional probability of e. So, we assign the same probability on the condition that either h_{H} or h_{VH} is true. We can formally establish this as follows. First, h_{H} and h_{VH} are incompatible, so that $P(h_{\mathrm{H}} \vee h_{\mathrm{VH}}) = P(h_{\mathrm{H}}) + P(h_{\mathrm{VH}})$ and $P(h_{\mathrm{H}} \vee h_{\mathrm{VH}} \mid e) = P(h_{\mathrm{H}} \mid e) + P(h_{\mathrm{VH}} \mid e)$. Also, $P(e \mid h_{\mathrm{H}}) = P(e \mid h_{\mathrm{VH}})$ because h_{H} and h_{VH} are empirically equivalent. We can then apply Bayes's Theorem to $P(e \mid h_{\mathrm{H}} \vee h_{\mathrm{VH}})$ and plug in these equations, as follows:

$$\begin{aligned}
P(e \mid h_{\mathrm{H}} \vee h_{\mathrm{VH}}) &= \frac{P(h_{\mathrm{H}} \vee h_{\mathrm{VH}} \mid e)P(e)}{P(h_{\mathrm{H}} \vee h_{\mathrm{VH}})} \\
&= \frac{\left[P(h_{\mathrm{H}} \mid e) + P(h_{\mathrm{VH}} \mid e)\right]P(e)}{P(h_{\mathrm{H}}) + P(h_{\mathrm{VH}})} \\
&= \frac{\left[\frac{P(e \mid h_{\mathrm{H}})P(h_{\mathrm{H}})}{P(e)} + \frac{P(e \mid h_{\mathrm{VH}})P(h_{\mathrm{VH}})}{P(e)}\right]P(e)}{P(h_{\mathrm{H}}) + P(h_{\mathrm{VH}})} \\
&= \frac{P(e \mid h_{\mathrm{H}})P(h_{\mathrm{H}}) + P(e \mid h_{\mathrm{VH}})P(h_{\mathrm{VH}})}{P(h_{\mathrm{H}}) + P(h_{\mathrm{VH}})} \\
&= \frac{P(e \mid h_{\mathrm{H}})\left[P(h_{\mathrm{H}}) + P(h_{\mathrm{VH}})\right]}{P(h_{\mathrm{H}}) + P(h_{\mathrm{VH}})} \\
&= P(e \mid h_{\mathrm{H}}).
\end{aligned}$$

To go back to the comparison of $J(h_H, e \mid b)$ and $J(h_H \vee h_{VH}, e \mid b)$, we now know that their respective numerators in the rewritten formula, $\log P(e \mid h_H) - \log P(e)$ and $\log P(e \mid h_H \vee h_{VH}) - \log P(e)$, are the same because $P(e \mid h_H)$ and $P(e \mid h_H \vee h_{VH})$ are the same. The relative epistemic worthiness of the two hypotheses, h_H and $h_H \vee h_{VH}$, therefore solely depends on the denominators, and we already know that $P(h_H) < P(h_H \vee h_{VH})$. The upshot is that $J(h_H, e \mid b) < J(h_H \vee h_{VH}, e \mid b)$. The disjunction of the hand hypothesis and the virtual hand hypothesis is epistemically more worthy than the hand hypothesis.

We can put the reasoning informally as follows. As explained in Chapter 4, the degree of epistemic worthiness $J(h, e \mid b)$ is the ratio of the earned information (its numerator) to the registered information (its denominator). So, if the hypotheses under consideration earn the same amount of information from the evidence, we should accept the one that registers the least amount of information. But when the hypotheses are empirically equivalent—as h_H, h_{VH} and $h_H \vee h_{VH}$ are—they earn the same amount of information from the evidence. So, we should accept the one that registers the least amount of information, which is $h_H \vee h_{VH}$ among the three.

The final issue of supersedure is whether there is justification for accepting h_H *after* the disjunction $h_H \vee h_{VH}$ is added to the body of beliefs b, which was initially empty. In other words, whether $J(h_H, e \mid h_H \vee h_{VH}) \geq t$. If so, h_H is not superseded by $h_H \vee h_{VH}$ because there is justification for accepting both $h_H \vee h_{VH}$ and h_H. It turns out, however, that $J(h_H, e \mid h_H \vee h_{VH}) = 0$. We can drive this from $P(e \mid h_H) = P(e \mid h_H \vee h_{VH})$ as follows:

$$J(h_H, e \mid h_H \vee h_{HV}) = \frac{\log P(e \mid h_H \wedge (h_H \vee h_{HV})) - \log P(e \mid h_H \vee h_{HV})}{-\log P(h_H)}$$
$$= \frac{\log P(e \mid h_H) - \log P(e \mid h_H \vee h_{HV})}{-\log P(h_H)}$$
$$= 0.$$

This means that once the disjunction $h_H \vee h_{VH}$ is accepted, there is no justification for adding h_H to it further. This makes good sense. Once the disjunction $h_H \vee h_{VH}$ is in the body of beliefs, h_H earns no more information from any evidence because it is empirically equivalent to $h_H \vee h_{VH}$. To conclude, the hand hypothesis h_H is superseded by the disjunction $h_H \vee h_{VH}$ of the hand hypothesis and the virtual hand hypothesis. More generally, although we have a strong case for accepting a particular hypothesis about the natural world in the presence of relevant evidence, it is superseded by the disjunction of the hypothesis and its counterpart proposed by the skeptic. We should therefore remain non-committal between any particular hypothesis about the natural world and its counterpart proposed by the skeptic.

5.5 Epistemic Justification up to Empirical Equivalence

It is straightforward to generalize the formal result obtained in the previous section to show that we cannot select and defend any hypothesis from a set of empirically equivalent rival hypotheses. Let $H = \{h_1, \ldots, h_n\}$ be a set of pairwise incompatible hypotheses. Assume that no member of H has been refuted with absolute certainty, and $h_1, \ldots, h_n$ are empirically equivalent in the sense that $P(e \mid h_1 \wedge b) = \ldots = P(e \mid h_n \wedge b)$ for any possible observation e. Let $H^* = \{h^*_1, \ldots, h^*_m\}$ be a proper subset of H, which could be a singleton with $m = 1$. Under these conditions, even if there is justification for accepting the disjunction $h^*_1 \vee \ldots \vee h^*_m$ of the members of H^* given some observation e, it is superseded by the disjunction $h_1 \vee \ldots \vee h_n$ of H given e. In other words, even if we select some members from H and find justification for their disjunction, it is always superseded by the disjunction of all members of H.

To establish the claim, it is sufficient to show that (a) $h_1 \vee \ldots \vee h_n$ is epistemically more worthy than $h^*_1 \vee \ldots \vee h^*_m$, and (b) once $h_1 \vee \ldots \vee h_n$ is accepted, $h^*_1 \vee \ldots \vee h^*_m$ has no epistemic worth at all. To state them formally, (a) $J(h^*_1 \vee \ldots \vee h^*_m, e \mid b) < J(h_1 \vee \ldots \vee h_n, e \mid b)$, and (b) $J(h^*_1 \vee \ldots \vee h^*_m, e \mid (h_1 \vee \ldots \vee h_n) \wedge b) = 0$. I show first that the disjunction $h_1 \vee \ldots \vee h_n$ is empirically equivalent to each disjunct, i.e. for any possible observation e, $P(e \mid (h_1 \vee \ldots \vee h_n) \wedge b) = P(e \mid h_1 \wedge b) = \ldots = P(e \mid h_n \wedge b)$. It suffices to show that $P(e \mid (h_1 \vee \ldots \vee h_n) \wedge b) = P(e \mid h_1 \wedge b)$ since $h_1, \ldots, h_n$ are empirically equivalent to each other.

$$P(e \mid (h_1 \vee \ldots \vee h_n) \wedge b) = \frac{P(h_1 \vee \ldots \vee h_n \mid e \wedge b)P(e \mid b)}{P(h_1 \vee \ldots \vee h_n \mid b)}$$

$$= \frac{[P(h_1 \mid e \wedge b) + \ldots + P(h_n \mid e \wedge b)]P(e \mid b)}{P(h_1 \mid b) + \ldots + P(h_n \mid b)}$$

$$= \frac{\left[\frac{P(e \mid h_1 \wedge b)P(h_1 \mid b)}{P(e \mid b)} + \ldots + \frac{P(e \mid h_n \wedge b)P(h_n \mid b)}{P(e \mid b)}\right]P(e \mid b)}{P(h_1 \mid b) + \ldots + P(h_n \mid b)}$$

$$= \frac{P(e \mid h_1 \wedge b)P(h_1 \mid b) + \ldots + P(e \mid h_n \wedge b)P(h_n \mid b)}{P(h_1 \mid b) + \ldots + P(h_n \mid b)}$$

$$= \frac{P(e \mid h_1 \wedge b)[P(h_1 \mid b) + \ldots + P(h_n \mid b)]}{P(h_1 \mid b) + \ldots + P(h_n \mid b)}$$

$$= P(e \mid h_1 \wedge b)$$

Similarly, $h^*_1 \vee \ldots \vee h^*_m$ is empirically equivalent to each disjunct. It follows that $h_1 \vee \ldots \vee h_n$ and $h^*_1 \vee \ldots \vee h^*_m$ are empirically equivalent to each other.

The empirical equivalence of the two disjunctions allows us to derive (a) $J(h^*_1 \vee \ldots \vee h^*_m, e \mid b) < J(h_1 \vee \ldots \vee h_n, e \mid b)$ by the following comparison.

$$J(h^*_1 \vee \ldots \vee h^*_m, e \mid b) = \frac{\log P(e \mid (h^*_1 \vee \ldots \vee h^*_m) \wedge b) - \log P(e \mid b)}{-\log P(h^*_1 \vee \ldots \vee h^*_m \mid b)}$$

$$J(h_1 \vee \ldots \vee h_n, e \mid b) = \frac{\log P(e \mid (h_1 \vee \ldots \vee h_n) \wedge b) - \log P(e \mid b)}{-\log P(h_1 \vee \ldots \vee h_n \mid b)}$$

The comparison is straightforward since their numerators are the same from the empirical equivalence of the two disjunctions. It follows from $P(h^*_1 \vee \ldots \vee h^*_m \mid b) < P(h_1 \vee \ldots \vee h_n \mid b)$ that $-\log P(h^*_1 \vee \ldots \vee h^*_m \mid b) > -\log P(h_1 \vee \ldots \vee h_n \mid b)$, and hence $J(h^*_1 \vee \ldots \vee h^*_m, e \mid b) < J(h_1 \vee \ldots \vee h_n, e \mid b)$. The remaining task is to establish (b) $J(h^*_1 \vee \ldots \vee h^*_m, e \mid (h_1 \vee \ldots \vee h_n) \wedge b) = 0$, which is also straightforward:

$$\begin{aligned} &J(h^*_1 \vee \ldots \vee h^*_m, e \mid (h_1 \vee \ldots \vee h_n) \wedge b) \\ &= \frac{\log P(e \mid (h^*_1 \vee \ldots \vee h^*_m) \wedge (h_1 \vee \ldots \vee h_n) \wedge b) - \log P(e \mid (h_1 \vee \ldots \vee h_n) \wedge b)}{-\log P(h^*_1 \vee \ldots \vee h^*_m \mid (h_1 \vee \ldots \vee h_n) \wedge b)} \\ &= \frac{\log P(e \mid (h^*_1 \vee \ldots \vee h^*_m) \wedge b) - \log P(e \mid (h_1 \vee \ldots \vee h_n) \wedge b)}{-\log P(h^*_1 \vee \ldots \vee h^*_m \mid (h_1 \vee \ldots \vee h_n) \wedge b)} \\ &= 0. \end{aligned}$$

The conclusion from (a) and (b) is that the disjunction of the members of any proper subset of *H* is superseded by the disjunction of all members of *H*.

The best epistemic practice is therefore to accept the disjunction of all empirically equivalent hypotheses and remain non-committal among them. The reason is not that all empirically equivalent rival hypotheses are of equal epistemic worthiness. Some of them may be epistemically more worthy than others, and there may even be prima facie justification for accepting some of them, but they are always superseded by the disjunction of all empirically equivalent hypotheses. We should therefore only aim at epistemic justification up to empirical equivalence, and abandon the unachievable goal of selecting and defending any particular hypothesis among empirically equivalent rivals. This is a significant retreat, but it is not the first time in our intellectual history that we are forced to abandon an unachievable goal and retreat to a less ambitious goal. Many epistemologists of the past strived for the Cartesian goal of absolute certainty, but we have long abandoned it in favor of probabilistic justification at least for empirical statements. The dual-component format of epistemic justification, which is superior to the Lockean format, reveals that we should retreat further to the goal of epistemic justification up to empirical equivalence. The remainder of the section examines implications of the retreat.

The first thing to note is that despite its radical appearance, the impact of the retreat to epistemic justification up to empirical equivalence is actually limited. This is because empirically equivalent hypotheses assign the same probability to any possible observation. For example, even if we are BIVs, we will notice any difference because we will have exactly the same sensory experience as we would in the natural world. Our expectations of the future experiences are exactly the same regardless of which hypothesis is true. So, if we remain non-committal and only accept the disjunction of empirically equivalent hypotheses, our expectations of the future experiences are still exactly the same. The retreat to epistemic justification up to empirical equivalence therefore makes no difference as far as our expectation of the experience is concerned. There are, of course, psychological impacts. It may make some difference—perhaps even a major difference—in some people's mind if they can no longer accept that they live in the natural world. But they should come to terms with the limit of our epistemic reach, just as we did in abandoning the Cartesian goal of absolute certainty. In case the negative psychological impacts are disruptive in some people's lives, they can hold faith-based beliefs in the natural world that are not based on evidence. As discussed in the previous chapter, it is harmless to hold faith-based beliefs for pragmatic reasons so long as they are kept apart from the body of beliefs that serves as the general basis of deliberations.

The retreat to epistemic justification up to empirical equivalent may have greater impacts on moral deliberations. Take hedonistic utilitarianism. In its simple version, the moral evaluation of actions are grounded in the amounts of pleasures and pains that they bring about in all sentient beings. Estimating the amounts is a daunting task, but the task becomes even more complicated if we remain non-committal between the natural world hypothesis and the BIV hypothesis. Presumably, the apparent dogs and cats that populate the virtual world in the BIV scenario feel no pleasure or pain. So, only the pleasures and pains experienced by humans (human brains in vats) count in the BIV scenario. Moreover, the particular BIV scenario we have been referring to for the purpose of illustration is only one version among many that are empirically equivalent to the natural world hypothesis. For example, there may be only one sentient being (one brain in a vat) if the solipsist version of the BIV hypothesis is true. Hedonistic utilitarianism collapses into hedonistic egoism in that scenario. The solipsist BIV scenario also complicates moral deliberations in the Kantian tradition. Do virtual humans in the scenario count as rational beings to which moral commands apply?

It may be pointed out that these are not new problems due to Cartesian skepticism. Even if we can somehow solve Cartesian skepticism in favor of the natural world hypothesis, we still face a similar challenge with regard to other minds, i.e. there may only be a single mind—your own—while all other human-like beings are zombies who behave like humans but lack any conscious experience. This scenario also affects moral deliberations of hedonistic utilitarianism, Kantian deontology, etc. in the same way the solipsist BIV scenario does. However, the problem of other minds is itself a problem of empirically equivalent alternative hypotheses. The big picture is that the problem of empirically equivalent hypotheses—whether it is in the form of Cartesian skepticism or the problem of other minds—affects moral deliberations, and epistemic justification up to empirical equivalence solves these problems in favor of the skeptic. We have to take into account all empirically equivalent hypotheses, and that complicates moral deliberations.

5.6 Epistemic Justification up to Equal Likelihood

Some people may suspect that embracing Cartesian skepticism has its greatest impact on religious beliefs. If the BIV hypothesis is true, then our brain may not be destroyed at our "death" in the virtual world, but is recycled to go through another round of life, perhaps in a virtual heaven.[19] Belief in an afterlife is difficult to square with the natural world hypothesis, but it seems to fit into the BIV hypothesis easily except that the afterlife may only take place in the virtual world. Of course, it is hard to find any empirical *support* for the BIV scenario with an afterlife, but there seems to be no justification for denying it either if we are to remain non-committal between the natural world hypothesis and the BIV hypothesis. Should we then withhold judgment about the possibility of an afterlife? On some reflection, however, the answer to the question seems to be no because the BIV scenario with an afterlife is not empirically equivalent to the natural world hypothesis. The scenario predicts the occurrence of posthumous sensory experience, while the natural world hypothesis makes no such prediction. It looks like we need not withhold judgment between them even if we retreat to epistemic justification up to empirical equivalence. It turns out, however, that the difference in their predictions is not good reason to favor the natural world hypothesis over the BIV scenario with an afterlife. Instead, it points to a much more disturbing consequence of the dual-component format than epistemic justification up to empirical equivalence.

The disturbing consequence is epistemic justification up to *equal likelihood*. Here is the idea. Let h_{BIV+} be the BIV scenario with an afterlife. The natural world hypothesis h_{NW} and the hypothesis h_{BIV+} are not empirically equivalent, i.e. $P(e \mid h_{NW} \wedge b) \neq P(e \mid h_{BIV+} \wedge b)$ for some possible evidence e. However, it is still the case that $P(e_A \mid h_{NW} \wedge b) = P(e_A \mid h_{BIV+} \wedge b)$ for the body of *actual* evidence e_A obtained by the subject so far, who we assume is not dead yet. Let us call hypotheses $h_1, \ldots, h_n$ *equally likely* when $P(e_A \mid h_1 \wedge b) = \ldots = P(e_A \mid h_n \wedge b)$ for the body of *actual* evidence e_A.[20] It was shown in Section 4 that if all of the members $h_1, \ldots, h_n$ of the set H are empirically equivalent, pairwise incompatible, and none of them is refuted with absolute certainty, then the disjunction $h^*_1 \vee \ldots \vee h^*_m$ of any proper subset H^* of H is superseded by the disjunction $h_1 \vee \ldots \vee h_n$. I assumed there that $h_1, \ldots, h_n$ are empirically equivalent because the subject of interest at that point was Cartesian skepticism, where the rival hypotheses are empirically equivalent. However, the same formal proof establishes that if all of the members $h_1, \ldots, h_n$ of the set H are *equally likely*, pairwise incompatible, and none of them is refuted with absolute certainty, then the disjunction of any proper subset H is superseded by the disjunction of all of the members of H.

First, if $h_1, \ldots, h_n$ are equally likely, the disjunction $h_1 \vee \ldots \vee h_n$ is also as likely as each disjunct is. Similarly, if $h^*_1, \ldots, h^*_m$ are equally likely, the disjunction $h^*_1 \vee \ldots \vee h^*_m$ is also as likely as each disjunct is. So, the two disjunctions $h_1 \vee \ldots \vee h_n$ and $h^*_1 \vee \ldots \vee h^*_m$ are equally likely. We can use this result to show that (a) $J(h^*_1 \vee \ldots \vee h^*_m, e_A \mid b) < J(h_1 \vee \ldots \vee h_n, e_A \mid b)$, and (b) $J(h^*_1 \vee \ldots \vee h^*_m, e_A \mid (h_1 \vee \ldots \vee h_n) \wedge b) = 0$. So, the disjunction of the members of any proper subset H^* of H is superseded by the disjunction of all the members of H. The formal proof is exactly the same as before, except that e_A replaces e. The upshot is that we cannot select and defend any hypothesis from a set of *equally likely* rival hypotheses. If $h_1, \ldots, h_n$ are equally likely, we should simply accept their disjunction. So, we should remain non-committal between the natural world hypothesis and the BIV scenario with an afterlife. More generally, we should only aim at *epistemic justification up to equal likelihood*, and abandon the unachievable goal of selecting and defending any particular hypothesis among equally likely rivals.[21]

To clarify again, the consequence is not that all hypotheses that are equally likely are epistemically worthy to the same degree. That is not the formal result. Some of the equally likely hypotheses can be epistemically more worthy than others. The formal result is that their disjunction is always epistemically more worthy than any of its disjuncts, and that once the disjunction is accepted, none of its disjuncts has any epistemic worthiness at all. Epistemic justification up to equal likelihood therefore suggests that we partition all hypotheses by their likelihoods with regard to the body of *actual* evidence e_A, and combine those hypotheses that are equally likely in the form of a disjunction. We then proceed to evaluate only the resulting disjunctions with no attempt to select and defend any of their disjuncts.

The retreat to epistemic justification up to equal likelihood has far more serious implications than does the retreat to epistemic justification up to empirical equivalence. This is because many equally likely hypotheses, such as the natural world hypothesis and the BIV scenario with an afterlife, make different predictions. We can no longer say that our expectations of the future experiences are the same regardless of which hypothesis is true. Nor can we say that our expectations of the future experiences are the same whether we accept one of the hypotheses or accept only their disjunction. Instead we are urged to remain non-committal between hypotheses that make different predictions, such as h_{NW} and h_{BIV+}, as long as they are equally likely with regard to the body of actual evidence e_A. This is devastating for the practice of induction. For example, the body of actual evidence e_A obtained so far supports the hypothesis h_S that the sun rises in the east in the morning, and we predict o_S that the sun will rise in the east tomorrow morning. The reasoning seems compelling even if we set aside the broader theories in astronomy that make the same prediction. However, with regard to the body of actual evidence e_A obtained so far, the hypothesis h_S is as likely as the alternative hypothesis h_S* that the sun rises in the east in the morning till today but rises in the west in the morning thereafter. The dual-component format of epistemic evaluation therefore urges us to remain on-committal between h_S and h_S* because h_S is superseded by their disjunction $h_S \vee h_S$*. We cannot incorporate h_S as the basis of future deliberation.[22]

Two clarifications are in order. First, it may actually seem reasonable to remain non-committal between h_S an h_S* because we cannot rule out h_S* that is consistent with the past observation. That is true, but the dual-component format also applies to equally likely *probabilistic* hypotheses. Consider, for example, the hypothesis h_{PS} that it is highly probable that the sun rises in the east in the morning, and the hypothesis h_{PS}* that it is highly probable that the sun rose in the east in the morning till today but rises in the west in the morning thereafter. If we only aim at epistemic justification up to equal likelihood, as recommended by the dual-component format, we should remain non-committal between h_{PS} and h_{PS}*. We should only accept their disjunction $h_{PS} \vee h_{PS}$*, instead of accepting h_{PS} as the basis of future deliberation. This is so despite the repeated past observations of the sunrise in the east.

The second clarification concerns the relation between the problem of induction due to equal likelihood and Goodman's new riddle of induction (Goodman 1955). They look somewhat similar, but they are actually orthogonal to each other. In the new riddle of induction, the challenge arises in the form of an alternative predicate, such as "grue", grounded in an alternative conceptual space.[23] Even if we solve the new riddle of induction by successfully defending our current conceptual space in which grue is not a natural property, the problem of equal likelihood persists since the choice of a conceptual space plays no role in it. It does not matter for the purpose of equal likelihood that the alternative hypothesis h_S* is complex and contrived in a given conceptual space, as long as it is as likely as the hypothesis h_S is. Meanwhile, even if we solve the problem of equal likelihood in induction relative to our current conceptual space, the new riddle of induction about an alternative conceptual space remains. My concern in this book is the problem of induction due to equal likelihood.

To summarize the main results of the chapter, the dual-component format of epistemic evaluation recommends that we only aim at epistemic justification up to empirical equivalence, and abandon the goal of selecting and defending a particular hypothesis among empirically equivalent rivals. This means that we must come to terms with Cartesian skepticism. This may be possible, but the dual-component format of epistemic evaluation also recommends that we only aim at epistemic justification up to equal likelihood, and abandon the goal of selecting and defending a particular hypothesis among equally likely rivals. This means that we must come to terms with inductive skepticism. This is much harder.

Some people may take this to be sufficient reason for rejecting the dual-component format of epistemic evaluation. There are, however, compelling reasons to adopt the dual-component format, as we discussed in Chapter 4. It is superior to the Lockean format, and there are no other formats on offer that are equally compelling. So, I stand by the dual-component format of epistemic evaluation until a better alternative is found. If we retain the dual-component format, one possible way of avoiding inductive skepticism is to regard induction as practical reasoning on which action to take, instead of epistemic reasoning on which hypothesis to accept. The dual-component format, which takes into account probabilities and the amounts of information, is intended for the latter, while the bases of practical reasoning are probabilities and utilities. The problem with this approach is that we would need to know relevant utilities that vary from one application to another.

I set aside these two responses—rejecting the dual-component format, or seeking pragmatic justification of induction—and pursue a third approach in the remainder of the book. Recall that there are two types of epistemic evaluation. One is qualitative and the other is quantitative. Qualitative epistemic evaluation determines whether or not we can accept a hypothesis in light of the evidence, while quantitative epistemic evaluation assigns a probability to the hypothesis. The dual-component format is a format of qualitative epistemic evaluation, and it makes use of probabilities that are assigned by quantitative epistemic evaluation. Quantitative evaluation is therefore prior to qualitative evaluation. In the next chapter I will propose and defend the relative-divergence format of epistemic evaluation for use in quantitative evaluation. My strategy against inductive skepticism is to eliminate equally likely alternative hypotheses at the stage of quantitative epistemic evaluation, so that we need not consider them at the stage of qualitative epistemic evaluation by the dual-component format.

Notes

1 The initial body of beliefs b is empty in the context of Cartesian skepticism. The body of evidence e consists of reflective sensory evidence, e.g. it appears to me visually that here is a hand.

2 I assume in the subsequent discussion that both the evaluator and the subject are aware of the logical entailment in question.

3 We can make the formulation of epistemic closure in the Lockean format stronger, viz. if $P(h_1 \mid e \wedge b) \geq t$ and $h_1 \wedge e \wedge b$ logically entails h_2 (even if h_1 by itself does not logically entail h_2), then $P(h_2 \mid e \wedge b) \geq t$.

4 In meliorative social epistemology, where the evaluator (the expert on the subject) endorses a statement to the subject (the non-expert member of the community), the evaluator does not take up everyday statements, such as that here is a hand, but the evaluator can endorse a relevant epistemic guideline to the subject, e.g. that the subject can accept a particular statement h about the natural world in the presence of sensory experience whose representational content is that h unless she notices something suspicious around her.

5 It does not count if there happens to be a hand in front of the brain in a vat. The hand is not "here" in the relevant sense (not in front of the visual sensor through

which the subject is understood to have received the visual input) because there is no visual sensor connected to the BIV.

6 Violating epistemic closure is "abominable" to use DeRose's (1995) word.

7 Unless, of course, the concept of epistemic justification itself is analyzed analogously to the concept of knowledge.

8 This is also true of multiple-premise logical entailment. If we have already added all of the statements $h_1, \ldots, h_n$ to the body of beliefs b, and $h_1, \ldots, h_n$ together with b logically entails h_{n+1}, then we can add h_{n+1} to $b \wedge h_1 \wedge \ldots \wedge h_n$ by analytic expansion that is free of risk and devoid of additional information.

9 A stronger version of epistemic closure under logical equivalence also holds, viz. if $J(h_1, e \mid b) \geq t$ and h_1 and h_2 are logically equivalent on condition of b (even if they are not logically equivalent unconditionally), then $J(h_2 \mid e \wedge b) \geq t$ for any threshold t.

10 For example, it holds trivially in the Lockean format because if h_1 and h_2 are logically equivalent, then $P(h_1 \mid e \wedge b) = P(h_2 \mid e \wedge b)$. It follows immediately that $P(h_1 \mid e \wedge b) \geq t$ if and only if $P(h_2 \mid e \wedge b) \geq t$ for any threshold t. A stronger version of epistemic closure under logical equivalence also holds in the Lockean format, viz. if $P(h_1 \mid e \wedge b) \geq t$ and h_1 and h_2 are logically equivalent on condition of $e \wedge b$ (even if they are not logically equivalent unconditionally or on condition of b alone), then $P(h_2 \mid e \wedge b) \geq t$ for any threshold t.

11 See Shogenji (2012) and Moretti and Shogenji (2017).

12 The analysis of the Linda case proposed here resembles the confirmation analysis of the case (Sides et al. 2002; Crupi, Fitelson and Tentori 2008). This is not surprising. As noted in Chapter 4, the measure J meets the basic requirements of a confirmation measure, though it is conceptually very different from the measures of confirmation. See Moretti and Shogenji (2017) for a comparison of the confirmation analysis and the dual-component analysis of the Linda case.

13 The hand hypothesis h_{H} has the skeptical counterpart, the virtual hand hypothesis h_{VH}, that the supercomputer sends the BIV the sensory input of apparently seeing a hand in front of the body. The evidence e of apparently seeing a hand in front of the body does not favor h_{H} over h_{VH}, but that is consistent with $P(h_{\mathrm{H}} \mid e) > P(h_{\mathrm{H}})$, which follows from $P(e \mid h_{\mathrm{H}}) > P(e \mid \neg h_{\mathrm{H}})$ by PESC of Chapter 3. Note that the hypothesis $\neg h_{\mathrm{H}}$ comprises not only h_{VH} but also many hypotheses that are not empirically equivalent to h_{H}.

14 It does not matter for the present purpose whether e' is a "rebutting" defeater or an "undercutting" defeater (Pollock and Cruz 1999, Ch. 7).

15 If the original set of hypotheses is $\{h_1, \ldots, h_n, h_{\mathrm{C}}\}$ where h_{C} is the catch-all (none of the above) hypothesis, then the formulation of a promising novel hypothesis h_{n+1} can increase the probability of h_{C} to which it belongs, and reduce the probabilities of the disjunction $h_1 \vee \ldots \vee h_n$ of the previously known hypotheses.

16 As noted in the previous chapter, the threshold of sufficient epistemic worthiness t should be set above zero, so that a hypothesis for which none of the registered information is earned would not be considered sufficiently epistemically worthy. This has the consequence that it is impossible for both h and its negation $\neg h$ to be sufficiently epistemically worthy because $J(h, e \mid b) \geq t > 0$ only if its numerator $\log P(h \mid e \wedge b) - \log P(h \mid b)$ is positive, i.e. only if $P(h \mid e \wedge b) > P(h \mid b)$. This makes it impossible that $P(\neg h \mid e \wedge b) > P(\neg h \mid b)$, and thus impossible that $J(\neg h, e \mid b) \geq t > 0$. However, making it impossible for both h and its negation $\neg h$ to be sufficiently epistemically worthy is not the reason for setting t above zero. Even if t is set at zero or below zero, once h or $\neg h$ is added to b, there is no justification for adding the other.

17 This point only applies to the original version of the lottery paradox. As shown in Chapter 4, the Lockean format is unable to handle the variant of the lottery

paradox in which the statements $p_1, \ldots, p_m$ are not mutually exclusive, but probabilistically independent of each other.

18 It may appear strange that the degree of epistemic worthiness is a direct function of the prior probability here. The degree of epistemic worthiness, which is a direct function of the amount of information, should be an inverse function of the prior probability since the amount of information is an inverse function of the prior probability. There is, however, no contradiction. When the conditional probability $P(h \mid e)$ is fixed, the degree of epistemic worthiness is an inverse function of the prior probability. However, when the numerator of $J(h, e \mid b)$ is fixed, of which the prior probability is also a component, the degree of epistemic worthiness is a direct function of the prior probability.

19 Bostrom (2003) raises this point with regard to computer simulations of the world, though simulation he considers does not involve brains in vats.

20 The term "likely" in "equally likely" is derived from the term "likelihood" in the technical sense, and is distinct from "probable". In most cases two hypotheses that are equally likely are not equally probable, i.e. it is often the case that $P(e_A \mid h_1 \wedge b) = P(e_A \mid h_2 \wedge b)$ but $P(h_1 \mid e_A \wedge b) \neq P(h_2 \mid e_A \wedge b)$.

21 The natural world hypothesis h_{NW} and the BIV hypothesis with an afterlife h_{BIV+} are empirically equivalent up to death in the sense that $P(e \mid h_{NW}) = P(e \mid h_{NW})$ for any possible evidence e up to death, but equal likelihood is much less demanding, i.e. it only requires that $P(e_A \mid h_{NW}) = P(e_A \mid h_{NW})$ for the body of *actual* evidence e_A obtained so far.

22 It cannot be ruled out that there may be justification for accepting the prediction o_S (that the sun will rise in the east tomorrow morning) directly from e_A with no mediation of h_S. It is hard, however, to even imagine a body of beliefs that consists of only disjunctions of equally likely hypotheses, and still allows justification of o_S from e_A.

23 See Gärdenfors (2000) for a lucid account of the conceptual space and its implications for the problem of induction.

6 Divergence from the Truth

6.1 Probabilistic Models

I now turn my attention to quantitative epistemic evaluation and introduce a new format of epistemic evaluation intended for probabilistic models. The basic principles of probabilistic modeling this chapter expounds on are well understood among statisticians, especially in statistical learning theory.[1] I believe, however, that the distinctive features of probabilistic modeling and its relation to qualitative evaluation are not fully appreciated in the philosophical literature. So, I will review some basic points, starting with the notion of a model.

A model, as it is understood in the subsequent discussion, is a template that specifies the shape of a theory. For example, LIN (the linear model) and QUAD (the quadratic model) shown here specify shapes of a theory about the relation between x and y.

$$\text{LIN: } y = a_1 x + a_0$$
$$\text{QUAD: } y = a_2 x^2 + a_1 x + a_0$$

A model generates a theory (a fitted model) when we feed a sufficient amount of data to it. For example, two data points <–2, 2> and <1, 5> determine the parameter values $a_1 = 1$ and $a_0 = 4$ for LIN to generate the theory $y = x + 4$.[2] The generated theory is a statement, and is true if the stated relation holds between the two variables x and y. Otherwise, e.g. if the next data point is <–1, 1>, then the generated theory is false because $y \neq x + 4$ for $x = -1$ and $y = 1$. The QUAD model is more flexible than LIN and can accommodate these three data points <–2, 2>, <1, 5>, <–1, 1> by the theory $y = x^2 + 2x + 2$.

Though it is model-generated theories, and not models themselves, that are true or false, there is a sense in which a model is correct or incorrect, viz. a model is correct if the true theory is in the shape specified by the model, and incorrect if the true theory is not in the shape specified by the model. For example, if the true theory is $y = x^2 + 2x + 2$, then QUAD is correct while LIN is incorrect. If the true theory is $y = x + 4$, then LIN is correct.

Note that QUAD is also correct in this case because we can set the parameter $a_2 = 0$ to reduce QUAD to LIN.[3] In general, if a polynomial model of some degree is correct, then any polynomial model of a higher degree is also correct, where POLY(n) below is the polynomial model of degree n:

$$\text{POLY}(n)\text{: } y = \sum\nolimits_{i=0}^{n} a_i x^i$$

We obtain LIN for $n = 1$ and QUAD for $n = 2$. If we want our model to be correct, and that is our only goal, then it is better to set the degree of the polynomial model higher. More generally, a model of greater complexity (greater flexibility) has a better chance of being correct. However, it is not sensible to choose a polynomial model of a very high degree if there are only a small number of data points. This is because the model would not generate any theory from the data. For example, if there are only two data points, a polynomial model of the degree two (QUAD) or higher fails to generate any theory because there are many combinations of parameter values that are consistent with the data.

In some cases a model fails to generate a theory because we have no values in the data set for some variables. For example, the model LIN2 with two independent variables x_1 and x_2 does not generate any theory if no value of x_2 is available.

$$\text{LIN2: } y = a_2 x_2 + a_1 x_1 + a_0$$

Of course, we can set the parameter value $a_2 = 0$ to make the value of x_2 irrelevant. The model can then generate a theory without the value of x_2. However, setting $a_2 = 0$ with no data to support it amounts to abandoning LIN2 in favor of the simpler model LIN: $y = a_1 x_1 + a_0$. The point remains that LIN2 is inappropriate when no value of x_2 is available. Either way, i.e. due to a small number of data points or due to the absence of required information in the data, if a model does not generate a theory from the data, the model is too complex for the given data. It is only when the data becomes sufficiently rich that such a model should be considered. This is an obvious point, but I put a special emphasis on it because it plays an important role in the next chapter: If a model fails to generate a theory from the given data, we should reject it in favor of a simpler model that does.

The relative-divergence format of epistemic evaluation I propose in this chapter is intended for models, but not just any models. It evaluates *probabilistic* models, whose connection to the truth is even weaker. A probabilistic model generates a theory that relates variables only probabilistically, or in the case of a constant function it generates a theory that states the outcome only probabilistically. To picture it in a concrete way, consider the following unconventional slot machine. When you insert a dollar, the machine displays its default color of green (G). Then, each time you press the button, it displays either green (G) or blue (B). Suppose the sequence so far is [G]GBGBBBGGBG, of which the first [G] is the default color that does not count. I call this data sequence "DS". An obvious model to consider for this machine is BNL (the Bernoulli distribution) that each time you press the button, the machine displays G with some probability p.[4]

$$\text{BNL: } P(\text{G}) = p$$

BNL can generate a deterministic theory with $p = 0$ or $p = 1$, but DS has already eliminated that possibility because DS contains both Gs and Bs. The theory that BNL generates in light of DS is therefore probabilistic with the value of p between zero and one, but what is the most appropriate value?

It is sensible to choose $p = 0.5$ in this case because DS consists of five Gs and five Bs. Indeed, that is the theory that BNL generates by the method of maximum likelihood (ML), where the parameter value that makes the given data most probable is chosen.[5] ML, or any statistical inference for that matter, does not ensure the truth of the generated theory. It is possible, for example, that the true parameter value is $p = 0.4$ and the five-in-ten ratio of Gs in DS is due to random variation, or noise in the data. It is rather unusual that the ratio in a finite sample exactly matches the true probability. For example, even if $p = 0.5$ is the true probability, the probability that exactly five of ten outcomes are Gs is only about one quarter.[6] Still the parameter value $p = 0.5$ makes the observed data sequence DS more probable than does any other parameter value.[7] The point is that once we embrace a probabilistic model, we should expect the data to have some random variation, and this truism has an unsettling consequence: Probabilistic models usually generate false theories. We need to evaluate probabilistic models with this in mind.

BNL is a sensible probabilistic model for our slot machine, but we can think of more complex models such as MKV (the Markov chain).[8]

$$\begin{aligned} \text{MKV: } & P(\text{G}) = p \text{ if the previous color was G} \\ & P(\text{G}) = q \text{ if the previous color was B} \end{aligned}$$

According to MKV, the probability of getting a G depends on whether the immediately preceding outcome is a G or a B. Indeed, it looks like we get a G slightly less often after a G than after a B in DS. Take a closer look at the sequence DS: [G]GBGBBBGGBG. If we adopt MKV, it is sensible to choose $p = 0.4$ because out of five occasions where a G was the previous outcome, including the default G, the new outcome was a G twice. Similarly, it is sensible to choose $q = 0.6$ because out of five occasions where a B was the previous outcome (excluding the last B), the new outcome was a G three times. The combination of $p = 0.4$ and $q = 0.6$ is the theory that MKV generates by the method of maximum likelihood (ML).[9] So, MKV generates a different theory than does BNL when we feed the same data sequence DS to them.

With two parameters p and q to adjust, MKV is more complex than BNL, and BNL can be considered a special case of MKV for $p = q$. This is similar to the relation between QUAD and LIN, where LIN is a special case of QUAD for $a_2 = 0$. Just as QUAD is correct (the true theory is in the shape specified by it) if LIN is correct, MKV is correct if BNL is correct. So, if we want our model to be correct, and that is our only goal, then it is better to adopt MKV than to adopt BNL. However, as mentioned already, it is not sensible to choose a complex model if it does not generate a theory from the data. For example, MKV generates no theory given only the initial data [G]GB because we cannot determine the value of the parameter q. We should not adopt a model that fails to generate a theory from the given data. There is no difference in this regard between probabilistic and non-probabilistic models.

There is, however, a significant difference between probabilistic and non-probabilistic models due to random variation the former allows. In the case of QUAD and LIN, it is safer to adopt the more complex model QUAD as long as it generates a theory from the given data because there are no cases where QUAD generates a false theory while LIN generates the true theory. More generally, there are no cases where a polynomial model of a higher degree generates a false theory while a polynomial model of a lower degree generates the true theory. We cannot say the same for probabilistic models. For example, though BNL can be considered a special case of MKV for $p = q$, there are many cases where MKV generates a false theory from the given data while BNL generates the true theory from the same data. Suppose, for example, the theory that BNL generates from DS is the true theory. In other words, the slot machine displays G with the probability of 0.5 as the ratio in DS indicates. This makes BNL a correct model. MKV is also a correct model in the sense that the true theory with the parameter values $p = q = 0.5$ is in the shape specified by MKV. However, the theory that MKV actually generates from DS is false. This is because MKV chases random variation in DS, instead of closing in on the true theory, to set the two parameter values differently at $p = 0.4$ and $q = 0.6$.

Furthermore, even if MKV is correct while BNL is incorrect, it may still be better off to adopt BNL. Suppose the true theory T for the slot machine is $T(G) = 0.49$ after a G while $T(G) = 0.51$ after a B. MKV is then correct while BNL is incorrect since the probability of getting a G depends on the immediately preceding outcome. The theory that BNL generates for DS is therefore false, but the false theory $P(G) = 0.5$ barely misses the mark. Meanwhile, the theory that the correct model MKV actually generates from DS, viz. $P(G) = 0.4$ after a G and $P(G) = 0.6$ after a B, is more inaccurate. The reason for the greater inaccuracy is again random variation. The difference in the ratios of "G after G" vs. "G after B" in DS is largely noise in the data, and not a reflection of the underlying Markov chain. Additional flexibility allows the complex model to accommodate the data better, but that is a curse when the data is noisy.

We can draw some lessons from these observations. First, when it comes to a probabilistic model, it does not count much that the model we adopt is correct in the sense that the true theory is in the shape specified by it. It is not uncommon that a correct probabilistic model generates a false theory by chasing random variation in the data. Second, the crucial issue is not the probability with which the model generates the true theory. Probabilistic models usually generate false theories anyways. Third, and most importantly, some false theories are better than others, where the difference is their divergence from the true theory, or their degree of inaccuracy.[10] If the true theory T is $T(\text{G}) = 0.49$, then it is better for a model to generate the theory $P(\text{G}) = 0.5$ than the theory $P(\text{G}) = 0.6$ though both of them are false. In the special case where the true theory T is $T(\text{G}) = 1$, getting closer to the truth amounts to assigning a higher probability to G, but in general what counts is not a high probability, but the divergence from the true probability distribution, or the degree of inaccuracy.

I used simple examples, BLN and MKV, with only one binary independent variable (its value is either B or G), but the points hold for probabilistic models in general, viz. what is important is not whether the model is correct or incorrect, or whether the probability that the model generates the true theory is high or low. It is more important in selecting a probabilistic model that it generates a theory that is close to the truth. Take some probabilistic models with independent and dependent variables, e.g. the probabilistic versions of LIN and QUAD, where the relation between x (the independent variable) and y (the dependent variable) is not deterministic due to random variation.[11] The probabilistic QUAD with more parameters to adjust is more flexible and can accommodate the data better than does the probabilistic LIN with fewer parameters to adjust. Indeed, the probabilistic LIN can never fit the data better than the probabilistic QUAD because if the theory generated by the probabilistic LIN fits the data well, the probabilistic QUAD can set the parameter value at $a_2 = 0$ to match the probabilistic LIN's performance. However, the greater flexibility of the probabilistic QUAD also makes it more prone to chase random variation. Good fit with the data may or may not be an indication of closeness to the truth.

6.2 Measuring Inaccuracy

Since some assumptions that are valid in the evaluation of non-probabilistic models are not valid in the evaluation of probabilistic models, we need to be cautious when we evaluate the latter, where the key notion is closeness to the truth, or the degree of inaccuracy. There has been a growing interest in the concept of inaccuracy,[12] but there are still some basic points that are not widely recognized, especially with regard to measures of inaccuracy. In this section I lay some groundwork on the notion of inaccuracy with a focus on its measures, before taking up the challenge of evaluating probabilistic models in Section 3.

To begin with a simple case, suppose a probability is assigned to a singular statement, e.g. to the statement that the slot machine will display green when its button is pressed next time. Once we observe the actual outcome, we can evaluate the post hoc inaccuracy of the assigned probability. If the statement in question turns out to be true but the probability assigned to it is less than one, then the assignment is inaccurate to some degree since the most accurate probability for the true statement is one. We can also evaluate the degree of inaccuracy (the distance to the truth). The idea is straightforward: The further away the assigned probability is from one, the more inaccurate it is. The degree of inaccuracy is therefore inversely related to the probability assigned to the statement that turned out to be true. This doesn't say much since there are many inverse functions of the probability: $1 - P(\mathrm{G})$, $[1 - P(\mathrm{G})]^2$, $1/P(\mathrm{G})$, $-\log P(\mathrm{G})$, etc. They are all candidates of a measure of post hoc inaccuracy, or "a scoring rule" as it is called in the literature.[13]

Formally, a scoring rule $\mathrm{SR}(P; i)$ is defined for a probability distribution P over a partition $Y = \{y_1,\ldots, y_n\}$ of statements, and one member y_i of the partition. For example, the partition may consist of two members, green ($y_1 = \mathrm{G}$) and blue ($y_2 = \mathrm{B}$). Suppose one theory states that the probability distribution over this partition is $P(\mathrm{G}) = 0.3$ and $P(\mathrm{B}) = 0.7$. Depending on which member of the partition turns out to be true, a scoring rule $\mathrm{SR}(P; i)$ returns either the value $\mathrm{SR}(P; \mathrm{G})$ or the value $\mathrm{SR}(P; \mathrm{B})$. It is clear that $\mathrm{SR}(P; \mathrm{G})$ should be greater than $\mathrm{SR}(P; \mathrm{B})$ because $P(\mathrm{G}) = 0.3$ is further away from $T(\mathrm{G}) = 1$ than $P(\mathrm{B}) = 0.7$ is from $T(\mathrm{B}) = 1$. In other words, P is more inaccurate post hoc if G turns out to be true than if B turns out to be true.

A scoring rule measures the inaccuracy of a probability distribution relative to a particular outcome. The next step is to extend the notion to the inaccuracy of a probability distribution relative to the true probability distribution, i.e. how far the proposed probability distribution P is from the true probability distribution T. From now on I will call it "the divergence from the truth" to avoid confusion with the distance to the true outcome captured by a scoring rule. Suppose the true probability distribution T over the partition $\{\mathrm{G}, \mathrm{B}\}$ is $T(\mathrm{G}) = 0.4$ and $T(\mathrm{B}) = 0.6$. The distribution $P(\mathrm{G}) = 0.3$ and $P(\mathrm{B}) = 0.7$ is then inaccurate, but we may also want to evaluate the degree of inaccuracy. This is important for the purpose of comparison. Suppose, for example, that a rival theory Q states that the probability distribution over the same partition $\{\mathrm{G}, \mathrm{B}\}$ is $Q(\mathrm{G}) = Q(\mathrm{B}) = 0.5$. Both P and Q are inaccurate, but which of them is less inaccurate (closer to the truth)? A measure of the divergence from the truth should give an answer to the question.

In order to answer the question, it is helpful to consider, first, the average post hoc inaccuracy. When G turns out to be the outcome, *P* receives the score of SR(*P*; G) as its post hoc inaccuracy. Meanwhile, when B turns out to be the outcome, *P* receives the score of SR(*P*; B) as its post hoc inaccuracy. So, the average post hoc inaccuracy of *P* depends on how often G turns out to be the outcome, and how often B turns out to be the outcome. The answer is given by the true probability distribution *T*. For *T*(G) = 0.4 and *T*(B) = 0.6, the average post hoc inaccuracy of the probability distribution *P* is 0.4 × SR(*P*; G) + 0.6 × SR(*P*; B). Similarly, the average post hoc inaccuracy of the probability distribution *Q* is: 0.4 × SR(*Q*; G) + 0.6 × SR(*Q*; B). The general formula for the average post hoc inaccuracy of the probability distribution *P* over the partition $Y = \{y_1, \ldots, y_n\}$ is:

$$\sum_{i=1}^{n} T(y_i)\mathrm{SR}(P;i) \qquad (1)$$

where *T* is the true probability distribution over $Y = \{y_1, \ldots, y_n\}$.[14]

It may appear that the average post hoc inaccuracy of *P* is simply the ante hoc degree of inaccuracy of *P*, but it is not the degree of inaccuracy in the sense of the divergence from the truth. This becomes clear when we think of the average post hoc inaccuracy of the true probability distribution *T* itself.. In the present example, where the partition is {G, B}, the formula for the average post hoc inaccuracy of the true distribution *T* is: *T*(G)SR(*T*; G) + *T*(B)SR(*T*; B). To plug in the numbers, *T*(G) = 0.4 and *T*(B) = 0.6 are both positive, and SR(*T*; G) and SR(*T*; B) are also positive because *T*(G) = 0.4 < 1 and *T*(B) = 0.6 < 1. So, *T*(G)SR(*T*; G) + *T*(B)SR(*T*; B) as a whole is positive. This is not surprising because even if we identify and accept the true probability distribution *T*(G) = 0.4 and *T*(B) = 0.6, the actual outcome is still either G or B, and this makes even the true distribution post hoc inaccurate. So, the average post hoc inaccuracy of the true probability distribution *T* is not zero. There is nothing wrong about it, but the positive average post hoc inaccuracy of *T* makes it clear that we cannot regard the average post hoc inaccuracy as the degree of inaccuracy in the sense of the divergence from the true distribution *T*. Since the divergence of *T* from *T* itself is zero, the degree of inaccuracy of *T* in the sense of its divergence from the true distribution should be zero.

The problem is solved by normalizing the average post hoc inaccuracy, so that the inaccuracy of *T* itself is always zero.

$$\sum_{i=1}^{n} T(y_i)\mathrm{SR}(P;i) - \sum_{i=1}^{n} T(y_i)\mathrm{SR}(T;i) \qquad (2)$$

When the probability distribution *P* to evaluate is *T* itself, the formula makes the degree of inaccuracy trivially zero. To interpret formula (2) informally, it measures the divergence of P from T by an increase in the average post hoc inaccuracy that results from adopting P instead of T. This is the way the ante hoc inaccuracy of a probability distribution is understood in the subsequent discussion.

There is still one unresolved issue, the selection of a scoring rule, before we can judge which of the rival probability distributions is closer to the truth. Let us apply formula (2) to the probability distributions P and Q in the example:

$$[T(\mathrm{G})\mathrm{SR}(P;\mathrm{G}) + T(\mathrm{B})\mathrm{SR}(P;\mathrm{B})] - [T(\mathrm{G})\mathrm{SR}(T;\mathrm{G}) + T(\mathrm{B})\mathrm{SR}(T;\mathrm{B})]$$
$$[T(\mathrm{G})\mathrm{SR}(Q;\mathrm{G}) + T(\mathrm{B})\mathrm{SR}(Q;\mathrm{B})] - [T(\mathrm{G})\mathrm{SR}(T;\mathrm{G}) + T(\mathrm{B})\mathrm{SR}(T;\mathrm{B})]$$

For the purpose of comparison, we can ignore the identical second term and focus on the first term that measures the average post hoc inaccuracy. So, which of the two below is the greater given the numbers $T(\mathrm{G}) = 0.4$ and $T(\mathrm{B}) = 0.6$, $P(\mathrm{G}) = 0.3$ and $P(\mathrm{B}) = 0.7$, and $Q(\mathrm{G}) = Q(\mathrm{B}) = 0.5$?

$$0.4 \times \mathrm{SR}(P;\mathrm{G}) + 0.6 \times \mathrm{SR}(P;\mathrm{B})$$
$$0.4 \times \mathrm{SR}(Q;\mathrm{G}) + 0.6 \times \mathrm{SR}(Q;\mathrm{B})$$

We still cannot tell. We know that $\mathrm{SR}(P; \mathrm{G}) > \mathrm{SR}(Q; \mathrm{G})$ from $P(\mathrm{G}) < Q(\mathrm{G})$, and $\mathrm{SR}(Q; \mathrm{B}) > \mathrm{SR}(P; \mathrm{B})$ from $Q(\mathrm{B}) < P(\mathrm{B})$, because a scoring rule is an inverse function of the probability. But that is not enough. The answer depends on the choice of a particular scoring rule $\mathrm{SR}(P; i)$.

It is a major challenge to select a particular scoring rule among many inverse functions of the probability. To list some candidates:[15]

$$\mathrm{SR}_{\mathrm{C}}(P;i) =_{\mathrm{def}} 1 - P(y_i)$$
$$\mathrm{SR}_{\mathrm{SC}}(P;i) =_{\mathrm{def}} [1 - P(y_i)]^2$$
$$\mathrm{SR}_{\mathrm{R}}(P;i) =_{\mathrm{def}} \frac{1}{P(y_i)}$$
$$\mathrm{SR}_{\mathrm{L}}(P;i) =_{\mathrm{def}} \log\frac{1}{P(y_i)} = -\log P(y_i)$$

To resolve the issue, we need some additional formal constraint on a scoring rule beyond being an inverse function of the probability.

The key constraint comes from the role of formula (2). Since formula (2) measures the divergence of P from T, its value should never be negative and should be zero if and only if P is identical to T. This is equivalent to the constraint that formula (1) should take the minimum value if and only if P is identical to T because formula (1) is the first part of formula (2) and the second part of formula (2) does not depend on P. This constraint on the scoring rule is known as *Strict Propriety* in the literature.[16] It is a powerful constraint because among many inverse functions of the probability, only the logarithmic scoring rule $\mathrm{SR_L}(P; i) = -\log P(y_i)$ satisfies it (Bernardo 1970). It is for this reason that I adopt the logarithmic scoring rule as the official measure of post hoc inaccuracy.[17]

By spelling out formulas (1) and (2) with $\mathrm{SR_L}(P; i) = -\log P(y_i)$, we obtain the two measures below that are known, respectively, as the cross entropy $\mathrm{H}(T, P)$ and the Kullback-Leibler divergence $\mathrm{KL}(T \parallel P)$:[18]

$$\begin{aligned}
&\sum\nolimits_{i=1}^{n} T(y_i)\mathrm{SR_L}(P;i) \\
&\quad = \sum\nolimits_{i=1}^{n} T(y_i)\left[-\log P(y_i)\right] \\
&\quad = -\sum\nolimits_{i=1}^{n} T(y_i)\log P(y_i) \\
&\quad =_{\text{def}} \mathrm{H}(T,P)
\end{aligned} \tag{3}$$

$$\begin{aligned}
&\sum\nolimits_{i=1}^{n} T(y_i)\mathrm{SR_L}(P;i) - \sum\nolimits_{i=1}^{n} T(y_i)\mathrm{SR_L}(T;i) \\
&\quad = \sum\nolimits_{i=1}^{n} T(y_i)\left[-\log P(y_i)\right] - \sum\nolimits_{i=1}^{n} T(y_i)\left[-\log T(y_i)\right] \\
&\quad = \sum\nolimits_{i=1}^{n} T(y_i)\left[\log T(y_i) - \log P(y_i)\right] \\
&\quad = \sum\nolimits_{i=1}^{n} T(y_i)\log \frac{T(y_i)}{P(y_i)} \\
&\quad =_{\text{def}} \mathrm{KL}(T \parallel P)
\end{aligned} \tag{4}$$

When T and P are probability density functions of a continuous variable y, the cross entropy and the Kullback-Leibler divergence take the following forms:

$$\mathrm{H}(T,P) = -\int_{-\infty}^{\infty} T(y)\log P(y)dy \tag{5}$$

$$\mathrm{KL}(T \parallel P) = \int_{-\infty}^{\infty} T(y)\log \frac{T(y)}{P(y)} dy \tag{6}$$

Discrete or continuous, the cross entropy is our official measure of the average post hoc inaccuracy, and the Kullback-Leibler divergence is our official measure of the divergence from the true distribution.

We can now use these measures to answer various questions of inaccuracy. For example, it was asked earlier which of the two probability distributions P and Q is closer to the true probability distribution T, where $T(\mathrm{G}) = 0.4$

and $T(B) = 0.6$, $P(G) = 0.3$ and $P(B) = 0.7$, and $Q(G) = Q(B) = 0.5$. The Kullback-Leibler divergence gives the following answer:

$$\begin{aligned}\mathrm{KL}(T \,\|\, P) &= \sum\nolimits_{i=1}^{n} T(y_i)\log\frac{T(y_i)}{P(y_i)}\\ &= 0.4\times\log\frac{0.4}{0.3}+0.6\times\log\frac{0.6}{0.7}\\ &= \log(0.4/0.3)^{0.4}+\log(0.6/0.7)^{0.6}\\ &= \log\left[(0.4/0.3)^{0.4}\times(0.6/0.7)^{0.6}\right]\\ &\approx \log 1.0228\end{aligned}$$

$$\begin{aligned}\mathrm{KL}(T \,\|\, Q) &= \sum\nolimits_{i=1}^{n} T(y_i)\log\frac{T(y_i)}{Q(y_i)}\\ &= 0.4\times\log\frac{0.4}{0.5}+0.6\times\log\frac{0.6}{0.5}\\ &= \log(0.4/0.5)^{0.4}+\log(0.6/0.5)^{0.6}\\ &= \log\left[(0.4/0.5)^{0.4}\times(0.6/0.5)^{0.6}\right]\\ &\approx \log 1.0234\end{aligned}$$

We can see that P is slightly closer to T than Q is.[19]

6.3 The Relative-Divergence Format of Epistemic Evaluation

With formal measures $\mathrm{H}(T, P)$ and $\mathrm{KL}(T \,\|\, P)$ in hand, let us now return to the evaluation of probabilistic models. $\mathrm{H}(T, P)$ and $\mathrm{KL}(T \,\|\, P)$ are measures of the theory P, but we can also apply them to evaluate a probabilistic model that generates a theory from the data. A good probabilistic model is one that generates from the given data a theory that is close to the truth, i.e. a theory whose $\mathrm{KL}(T \,\|\, P)$ value is small. In order to state this idea formally, it is helpful to introduce some notation. When a model M generates a theory P from a data set D by the method of maximum likelihood (ML), I will write $\mathrm{KL}(T \,\|\, P)$ as $\mathrm{KL}(T \,\|\, \mathrm{ML}(M, D))$ to make the origin of the theory clear. So, given the data D and a list of candidate models $M_1,\ldots, M_n$, the best model is the one that generates from the data a theory whose divergence from the truth $\mathrm{KL}(T \,\|\, \mathrm{ML}(M_j, D))$ is the smallest among those generated by the candidate models. Note that if some models fail to generate a theory from the data, they are automatically disqualified. As noted earlier, models that generate no theory from the given data are too complex for the data.

I described the basic idea of model selection by the Kullback-Leibler divergence, but this is not enough for solving the problems of model selection. In normal cases of model selection we do not know the true theory T, and without knowing T we cannot calculate the Kullback-Leibler divergence $\mathrm{KL}(T \parallel \mathrm{ML}(M_j, D))$. To overcome this problem we need to take two further steps. The first step is to shift focus from the absolute divergence to the relative divergence: We can evaluate theories by their relative divergences from the truth (which theories are more/less divergent from the truth than others) without knowing their absolute divergence $\mathrm{KL}(T \parallel \mathrm{ML}(M_j, D))$ from the truth. Recall that we arrived at the Kullback-Leibler divergence by normalizing the average post hoc inaccuracy $\mathrm{H}(T, P)$, i.e. by subtracting the average post hoc inaccuracy of T itself from $\mathrm{H}(T, P)$, so that the divergence $\mathrm{KL}(T \parallel T)$ of T from itself will be trivially zero. So, when the theory P is generated by the model M_j from the data D, its divergence from the truth is the following difference:

$$\mathrm{KL}(T \parallel \mathrm{ML}(M_j, D)) = \mathrm{H}(T, \mathrm{ML}(M_j, D)) - \mathrm{H}(T, T)$$

Of the two terms on the right side, the subtrahend $\mathrm{H}(T, T)$ is independent of $\mathrm{ML}(M_j, D)$. Though the minuend $\mathrm{H}(T, \mathrm{ML}(M_j, D))$ is not the divergence from the truth, we can tell the relative divergences of the theories from the truth by comparing their $\mathrm{H}(T, \mathrm{ML}(M_j, D))$ values. For example, $\mathrm{KL}(T \parallel \mathrm{ML}(M_1, D)) - \mathrm{KL}(T \parallel \mathrm{ML}(M_2, D))$ amounts to $\mathrm{H}(T, \mathrm{ML}(M_1, D)) - \mathrm{H}(T, \mathrm{ML}(M_2, D))$:

$$\begin{aligned}
&\mathrm{KL}(T \parallel \mathrm{ML}(M_1, D)) - \mathrm{KL}(T \parallel \mathrm{ML}(M_2, D)) \\
&\quad = [\mathrm{H}(T, \mathrm{ML}(M_1, D)) - \mathrm{H}(T, T)] - [\mathrm{H}(T, \mathrm{ML}(M_2, D)) - \mathrm{H}(T, T)] \\
&\quad = \mathrm{H}(T, \mathrm{ML}(M_1, D)) - \mathrm{H}(T, \mathrm{ML}(M_2, D))
\end{aligned}$$

For the purpose of evaluating the relative divergences from the truth, we can ignore $\mathrm{H}(T, T)$ and focus on $\mathrm{H}(T, \mathrm{ML}(M_j, D))$.

We are not out of the woods yet because $\mathrm{H}(T, \mathrm{ML}(M_j, D))$ itself contains T and we cannot calculate it without knowing T. So, we need to take a further step, which is to estimate $\mathrm{H}(T, \mathrm{ML}(M_j, D))$ without knowing T. Fortunately, there are well-known methods of estimating $\mathrm{H}(T, \mathrm{ML}(M_j, D))$ in the absence of the knowledge of T. Among these methods, I discuss cross-validation (CV) in this section and Akaike information criterion (AIC) in Section 5. To estimate the inaccuracy of a theory that the model generates from the data, the cross-validation method partitions the given data into two subsets—the training data and the validation data—and lets the model generate a theory from the training data alone. We can then calculate the post hoc inaccuracy of the theory by comparing its prediction and the validation data. This process is repeated with different partitions of the data, and we obtain the cross-validation score of the model for the data by averaging the results. The cross-validation score is only an estimate of the model-generated theory's inaccuracy for the entire population because it is solely based on the given data. The model may perform worse for the entire population, but that is the nature of any statistical inference. We can only rely on the given data to estimate the best model for the entire population.[20] So, we select the model whose cross-validation score is the smallest.

There are different versions of cross-validation that are appropriate for different applications.[21] For example, in the *k*-fold version, the given data is divided into *k* equal-sized subsets, and each of the *k* subsets at a time is used as the validation data while the training data is the rest of the given data. There are further variations of *k*-fold cross-validation. For example, in "stratified" *k*-fold cross-validation, all of the *k* folds must have similar average outcomes. Meanwhile, in leave-*p*-out cross-validation, we leave *p* data points to use as the validation data, and use the rest of the given data as the training data. Leave-*p*-out cross-validation is not the same as *k*-fold cross-validation even when each of the *k* folds has *p* data points because there are more than *k* ways of selecting *p* data points from the data set. For example, there are $10!/[2! \times 8!] = 45$ ways of selecting two data points from the data set of size ten, so that we need to repeat the training-and-evaluating process forty-nine times, instead of five times required for *k*-fold cross-validation with $k = 5$. When *k* of *k*-fold cross-validation is the size *n* of the entire data set, one data point at a time is the validation data and the rest is the training data. This amounts to "leave-one-out cross-validation".

For an illustration, let us apply leave-one-out cross-validation to the model BNL and the data sequence DS: [G]GBGBBBGGBG. First, we remove one data point from DS and use the remainder as the training data. For example, if we remove the last data point and ignore the default [G] that plays no role in BNL, the training data is GBGBBBGGB. We then feed the training data to BNL to generate a theory *P* as usual, i.e. to determine the parameter value *p* by the method of maximum likelihood. Since four of the nine outcomes in the training data are Gs, the best parameter value is $p = 4/9$. We use this value to calculate the post hoc inaccuracy of the generated theory *P* for the validation data (the removed data point). Since the outcome in the validation data is a G, the score is:

$$\begin{aligned}\mathrm{SR_L}(P;\mathrm{G}) &= -\log P(\mathrm{G}) \\ &= -\log 4/9 \\ &= \log 9/4\end{aligned}$$

We repeat the same process for each of the ten data points, i.e. remove the data point, generate a theory based on the remaining data, and evaluate the generated theory by the removed data point. If the removed data point is a G, then we obtain the same parameter value and score as above, i.e. $p = 4/9$ and $\mathrm{SR_L}(P;\ \mathrm{G}) = \log 9/4$. If the removed data point is a B, then five of the nine outcomes in the training data are Gs. So, the best parameter value is $p = 5/9$ instead of 4/9, but the score turns out to be the same, since the predicted probability for the removed data point B is $1 - 5/9 = 4/9$:

$$\begin{aligned}\mathrm{SR_L}(P;\mathrm{B}) &= -\log P(\mathrm{B}) \\ &= -\log\,(1 - 5/9) \\ &= -\log 4/9 \\ &= \log 9/4\end{aligned}$$

After obtaining the score for each data point, we calculate their average to obtain the cross-validation score CV(M, D) for model M and data D, which is trivially CV(BNL, DS) = log 9/4 in the present case since the ten scores are identical.

As noted already, there are many versions of cross-validation, and cross-validation is not the only method of estimating the inaccuracy of the model-generated theory. There is no need to single out a particular method or a particular version of the method in exclusion of others. Leaving the method of estimation open, my proposal of the relative-divergence format of epistemic evaluation for probabilistic models is as follows:

> **The Relative-Divergence Format of Epistemic Evaluation:** Given the data D and a list of candidates $M_1,\ldots, M_n$ for a probabilistic model, select the model M_j such that the theory ML(M_j, D) that M_j generates from D has the smallest estimated divergence from the truth among the theories generated by the candidate models.

I will use this format of epistemic evaluation for probabilistic models to complement the dual-component format of epistemic evaluation that makes use of probabilities.

6.4 The Full Truth

I will examine some implications of the format in the next two sections, but I want to take a brief pause here to reflect on the notion of the true probability distribution. The relative-divergence format assumes that the truth relative to which the divergence of a proposed theory is measured is probabilistic, but is that plausible? It is likely that any theory we formulate under the resource constraints is only probabilistic for the purpose of making predictions in realistic situations. However, setting aside resource constraints, do we need to conceive of the truth itself probabilistically? If we keep adding independent variables to the model, all values of the dependent variables will eventually become determinate. To take an extreme case, if the exact spatiotemporal location of the event is included as an independent variable in the model, then at most a single event can satisfy any set of values for the independent variables, so that the values of all dependent variables are determined with the probability one.[22] Can we regard it as the full truth and measure the divergence of a proposed theory relative to the deterministic full truth?

Some people may like the deterministic understanding of the full truth for conceptual economy. If the true theory T is deterministic, then T is completely accurate, i.e. H(T, T) = 0. This makes KL(T || ML(M_j, D)) = H(T, ML(M_j, D)) – H(T, T) simply H(T, ML(M_j, D)). So, when we compare estimated inaccuracies of theories ML(M_j, D), we are actually comparing their estimated divergences from the full truth, and not their estimated relative divergences. The deterministic understanding of the full truth also makes the relation between truth and prediction more direct. Since the inaccuracy of a probabilistic theory is the average post hoc inaccuracy of its probabilistic prediction, if the divergence of the probabilistic theory from the truth is simply its inaccuracy, then its divergence from the truth is just the average post hoc inaccuracy of its probabilistic prediction.

However, despite all these apparent advantages, the deterministic understanding of the full truth is misguided—that is not the kind of truth we aim. Note first that since the exact spatiotemporal location singles out an event, no other independent variables are needed for the deterministic true theory. Given any spatiotemporal location, the deterministic true theory simply returns the actual values of any dependent variables for the event that is singed out. We may also have many such "full truths". For example, the sole independent variable in a theory of voting behavior in the presidential election in the US may be each voter's social security number, the spatiotemporal location of their birth, etc. Since each of them singles out a voter, there is a deterministic true theory of their voting behavior with any of them as the sole independent variable. There is no single theory that is *the* full truth. Further, and more importantly, in most of these full truths, the relation between the independent variable and the dependent variable is accidental. So, we can infer nothing about the population from the relation found in the sample data.

We can see why these are not the kind of truth we aim at by applying cross-validation to them with regard to the entire population (hypothetically). Take the true deterministic theory of voting behavior of which the sole independent variable is the voter's social security number. The theory returns one of the three voting values D, R or N for each social security number. If we remove one data point (one pair of a social security number and a voting value) to use as the validation data, the training data does not generate any theory. This is because in order to generate a theory from the entirely accidental relation, the model must determine the relation between the independent variable and the dependent variable for each pair of a social security number and a voting value. The model fails to determine the relation for any pair unless the pair is in the training data. Any such model is rejected by the relative-divergence format of epistemic evaluation because it is too complex to even apply cross-validation.

What is then the kind of truth we aim at? I answer the question in two steps. The first step is the local full truth. Consider all possible models M_j on a particular subject matter. Let each of them (hypothetically) generate a theory ML(M_j, *DP*) from the *full data DP* on the entire population instead of a sample data, and then estimate its relative divergence from the truth in the usual way by cross-validation. The theory generated by the model that performs the best is then the full true theory on the subject matter. The resulting theory is probabilistic in most cases and does not fit the data as well as a deterministic theory would, but that is the kind of theory we aim at on the subject matter.[23] Some people may point out that if the population of interest is finite and small, the best model may still pick up an accidental relation. The second part of my answer addresses this concern, viz. we aim at the global full truth of which the local full truths on particular subject matters are constituents. This does not simply mean that we investigate all subject matters. The local full truths on particular subject matters must fit together to form the global full truth. If the relation picked up by the best local model is only accidental, the theory will not fit well with the best theories on related subject matters.[24]

So, in all likelihood the full truth relative to which the estimated divergences of model-generated theories are compared is probabilistic. As a result, the full truth is inaccurate in that its average post hoc inaccuracy is positive. This means that the qualification "relative" remains in the relative-divergence format of epistemic evaluation. The inaccuracy $\mathrm{H}(T, \mathrm{ML}(M_j, D))$ of a theory generated by the model M_j from the data D is not its divergence from the truth. By comparing the inaccuracies $\mathrm{H}(T, \mathrm{ML}(M_j, D))$ of competing models, we can only tell their relative divergences from the full truth. The relation between the inaccuracy of the prediction and the divergence from the full truth is also indirect because the inaccuracy of its prediction is the inaccuracy of a theory, and not the divergence from the full truth.

I want to add one more remark on the relation between truth and prediction. It may be thought that $\mathrm{H}(T, \mathrm{ML}(M_j, D))$ can be more straightforwardly considered the inaccuracy of the prediction rather than the relative divergence from the truth. More generally, it may be thought that the real focus in model selection should be the predictive inaccuracy of the generated theory, and reducing the divergence from the truth is only a means for that goal. I want to point out in response that a model with the least relative divergence from the truth is not always a model with the least amount of inaccuracy in prediction, and there is good reason to choose the former when they are not the same. Here is an example to illustrate the point. Suppose there are only two models M_1 and M_2 under consideration, and they generate theories P and Q, respectively, from the data. Suppose further that M_1 performs better than M_2 in cross-validation, so that we should select M_1 by the relative-divergence format of epistemic evaluation. However, if the probabilities of the next outcome, say G, are very different between P and Q for some value of the independent variable x, some people may be reluctant to ignore the other model M_2 that generates Q. For example, if $P(\mathrm{G} \mid a) = 0.4$ while $Q(\mathrm{G} \mid a) = 0.9$ for the particular value a for x, it seems sensible to assign a probability somewhat higher than 0.4 in light of $Q(\mathrm{G} \mid a) = 0.9$ since it is possible that Q is closer to the truth.[25] After all, cross-validation is only an estimate and an estimate can miss the mark. Taking the inferior model into account is sensible if predicting the next outcome is the only concern. If we do it systematically, given any value $\alpha = x$, we may assign the probability $\omega[\mathrm{ML}(M_1, D)](\mathrm{G} \mid \alpha) + (1 - \omega)[\mathrm{ML}(M_2, D)](\mathrm{G} \mid \alpha)$ for some weight ω.[26]

However, the relative-divergence format of epistemic evaluation is intended for model selection, while no model under consideration generates a theory that assigns the probability $\omega[\mathrm{ML}(M_1, D)](\mathrm{G} \mid a) + (1 - \omega)[\mathrm{ML}(M_2, D)](\mathrm{G} \mid a)$ to the next outcome given $x = a$.[27] It is worth noting that predicting the next outcome is not the sole goal of selecting a model. The selected model will keep adjusting its parameters in response to additional data. Also, the selected model becomes the basis of epistemic evaluations beyond the immediate subject matter since many models on different subject matters can be combined to form a more comprehensive model. It is rather short-sighted to focus narrowly on the prediction of the next outcome. The relative-divergence format of epistemic evaluation aims at the quantitative alethic goal of making the probabilistic theory as close to the truth as possible, while the prediction of the next outcome is one of its many applications.

6.5 Simplicity and Accuracy

This section examines the relation between simplicity and accuracy. It is mentioned in Section 1 that when a simple probabilistic model is a special case of a complex probabilistic model, the theory generated by the simple model never fits the data better than does the theory generated by the complex model. For example, the theory generated by BNL never fits the data better than the theory generated by MKV because we can always set the parameters of MKV at $p = q$ to reduce it to BNL and make the theory fit the data at least equally well. However, it is also mentioned there that the theory generated by the simple model can be closer to the truth than is the theory generated by the complex model. This is shown in Section 1 by an example where the true theory is provided by stipulation. I now make the same point by the relative-divergence format of epistemic evaluation without stipulating the true theory T. More specifically, I show that with regard to the data sequence DS: [G]GBGBBBGGBG, the cross-validation score CV(BNL, DS) for BNL is smaller than the cross-validation score CV(MKV, DS) for MKV. This means that the theory ML(BNL, DS) generated by the simpler model BNL has a smaller estimated divergence from the truth than does the theory ML(MKV, DS) generated by the more complex model MKV.

Since it is already shown in Section 3 that CV(BNL, DS) = log 9/4, we only need to calculate CV(MKV, DS) for the comparison. There is some complication since MKV relates the probability of the outcome to the previous outcome. To take into account the role of the previous outcome, we need to regard each adjacent pair in DS as a data point. In other words, the relevant data set is {<[G], G>, <G, B>, <B, G>, <G, B>, <B, B>, <B, B>, <B, G>, <G, G>, <G, B>, <B, G>}. To apply cross-validation to this data set, we remove one data point from it, and use the rest of it as the training data. Suppose we remove the eighth data point. The training data is then {<[G], G>, <G, B>, <B, G>, <G, B>, <B, B>, <B, B>, <B, G>, <G, B>, <B, G>}. We then feed this training data to MKV to generate a theory, i.e. to determine the parameter values p and q by the method of maximum likelihood. Of the four pairs in the training data whose first member is a G, only one has a G as its second member. So, the first parameter value is $p = 1/4$. Of the five pairs in the training data whose first member is a B, three of them have a G as their second member. So, the second parameter value is $q = 3/5$. We use these parameter values to calculate the post hoc inaccuracy of the generated theory P for the validation data, which is the removed data point <G, G>. Since the first member of the validation data is a G, the relevant parameter is p, and the score for the outcome G is:

$$\begin{aligned} SR_L(P;G) &= -\log p \\ &= -\log 1/4 \\ &= \log 4 \end{aligned}$$

This is the post hoc inaccuracy of the theory that MKV generates from the training data when the eighth data point is removed.

We repeat the same process for each of the ten data points i.e. remove the data point, generate a theory from the remaining data, and evaluate the generated theory by the removed data point. If the removed data point is a <G, G>, then we obtain the same parameter values and the same score as above, viz. $p = 1/4$, $q = 3/5$ and $SR_L(P;\ G) = \log 4$. If the removed data point is a <G, B>, then the parameter values are $p = 1/2$ and $q = 3/5$, and the score is $SR_L(P;\ B) = -\log(1-1/2) = \log 2$. If the removed data point is a <B, G>, then the parameter values are $p = 2/5$ and $q = 1/2$, and the score is $SR_L(P;\ G) = -\log 1/2 = \log 2$. Finally, if the removed data point is a <B, B>, then the parameter values are $p = 2/5$ and $q = 3/4$, and the score is $SR_L(P;\ B) = -\log(1-3/4) = \log 4$. Since there are two <G, G>s, three <G, B>s, three <B, G>s and two <B, B>s in the original data set, the average post hoc score of the MKV for the data sequence DS is:

$$\begin{aligned} CV(MKV, DS) &= 0.2 \log 4 + 0.3 \log 2 + 0.3 \log 2 + 0.2 \log 4 \\ &= \log 4^{0.4} + \log 2^{0.6} \\ &= \log \left[2^{0.8} \times 2^{0.6}\right] \\ &\approx \log 2.639 \end{aligned}$$

We can see that CV(MKV, DS) is greater than CV(BNL, DS) = log 9/4 = log 2.25. In other words, we can estimate that the BNL-generated theory ML(BNL, DS) is closer to the true theory T than is the MKV-generated theory ML(MKV, DS). Cross-validation allows us to make this estimation solely on the basis of the available data DS with no additional information on the true theory T.

So, by the relative-divergence format of epistemic evaluation, we should select the simpler model BNL over the complex model MKV for the data sequence DS. It is worth reiterating that BNL can never accommodate the data better than MKV does. Indeed the MKV-generated theory ML(MKV, DS) fits DS better than the BNL-generated theory ML(BNL, DS) does. We can see this by comparing their "training errors" which are the average inaccuracies of ML(BNL, DS) and ML(MKV, DS) with regard to the *entire* data set DS with no separation between the training data and the validation data. The calculation of the training error is straightforward. To begin with BNL, since the parameter value of ML(BNL, DS) is $p = 0.5$, the score for a G is $SR_L(P; G) = -\log 0.5 = \log 2$, and the score for a B is also $SR_L(P; B) = -\log(1 - 0.5) = \log 2$. So, the average score is trivially log 2, and that is the training error of ML(BNL, DS). Meanwhile, the parameter values of ML(MKV, DS) are $p = 0.4$ and $q = 0.6$. Since the relevant parameter is $p = 0.4$ when the previous outcome is a G, the score for a G after a G is $SR_L(P; G) = -\log 0.4 = \log 5/2$, and the score for a B after a G is $SR_L(P; G) = -\log(1 - 0.4) = \log 5/3$. Since the relevant parameter is $q = 0.6$ when the previous outcome is a B, the score for a G after a B is $SR_L(P; G) = -\log 0.6 = \log 5/3$, and the score for a B after a B is $SR_L(P; B) = -\log(1 - 0.6) = \log 5/2$. Since there are two <G, G>s, three <G, B>s, three <B, G>s and two <B, B>s in the data set, the training error of ML(MKV, DS) is as follows:

$$
\begin{aligned}
0.2 \log 5/2 + 0.3 \log 5/3 + 0.3 \log 5/3 + 0.2 \log 5/2 \\
&= \log (5/2)^{0.4} + \log (5/3)^{0.6} \\
&= \log \left[(5/2)^{0.4} \times (5/3)^{0.6} \right] \\
&\approx \log 1.960
\end{aligned}
$$

The training errors confirm that the MKV-generated theory fits DS better, though it is the BNL-generated theory that is closer to the truth by the cross-validation estimate.

In short, the MKV-generated theory fits DS better than the BNL-generated theory does, which is expected because MKV is more flexible than BNL, but by the cross-validation estimate based on the same data DS, the BNL-generated theory is closer to the truth than is the MKV-generated theory. We can explain it in the following way: The MKV-generated theory fits the data better because MKV pushes the theory away from the true theory to accommodate random variation in DS. This is called "overfitting" in statistical modeling. Overfitting can occur for any probabilistic models, but complex models are more prone to this problem because complexity makes them more flexible to accommodate random variation. This is how the MKV-generated theory fits the data better than the BNL-generated theory does despite its greater estimated divergence from the truth.

The takeaway is that we cannot always regard a better fit with the data as an indication of a smaller divergence from the truth. We also need to consider the flexibility of the model that generates the theory. If the model is flexible, we need to mark down the generated theory to counterbalance the model's greater propensity for overfitting. In other words, we should select a flexible model over less flexible (simpler) rivals only if the theory generated by the model fits the data sufficiently better than those generated by less flexible (simpler) rivals. There are of course many clear cases of this kind, e.g. we should select MKV over BNL if the data sequence is [G]BGBGBGBGBG.[28] There are however many cases where it is not immediately clear whether the difference in the goodness of fit with the data is sufficiently great to counterbalance difference in the flexibility of the model.

Where there is doubt, we can always go back to cross-validation. However, the degree of fit (the training error) is much easier to calculate because there is no need to repeat the calculation with different divisions of the training data and the validation data as in cross-validation. So, it is of great practical value if there is a formal way of balancing the goodness of fit and the complexity of the model in estimating the relative divergence from the truth without performing cross-validation, especially when the data set is large and the candidate models are complex. The statistical science has made significant progress on the subject since a breakthrough in the mid 1970s, and it turns out the formal results are not only of great practical value, but theoretically illuminating as well.

The first and best known among these formal results is Akaike Information Criterion (AIC) due to Akaike (1974):

$$\text{AIC}: -\ln \text{L}(\text{ML}(M, D) \mid D) + K$$

Given the data *D* and a candidate model *M*, we can calculate its AIC score to estimate the relative divergence of the *M*-generated theory from the truth, where the divergence from the truth is measured by the Kullback-Leibler divergence. ML(*M*, *D*) is the theory that the model *M* generates from the data *D* by the method of maximum likelihood; L is the likelihood function, i.e. L(ML(*M*, *D*) | *D*) is the probability of the data *D* given the theory ML(*M*, *D*); and its negative natural logarithm – ln L(ML(*M*, *D*) | *D*) measures how inaccurate ML(*M*, *D*) is with regard to the data *D*. The second term *K* is the estimated amount of further inaccuracy of ML(*M*, *D*) with regard to the whole population, due to the complexity of the model *M*. Akaike has shown that when the negative natural logarithm is used for the first term, the amount of discount *K* for the estimated relative divergence is the number of adjustable parameters in the model.

I want to stress that AIC is an approximation. Akaike has shown it to be the best approximation under certain conditions for models that behave well, and there have since been many formulas similar to AIC that generalize Akaike's results.[29] So, in addition to cross-validation, we have a family of methods to estimate the relative divergence of a model-generated theory from the truth.[30] Unlike cross-validation, these methods produce good estimates only under certain conditions, but when these conditions are met, they are much easier to calculate and are theoretically illuminating. All information criteria that generalize Akaike's results share the basic structure with AIC: The first term measures the model's goodness of fit with the data, and the second term is the amount of discount due to the complexity (flexibility) of the model. In other words, AIC and its generalizations confirm analytically what we see in cross-validation scores, viz. though a complex (flexible) model can accommodate the data better than its simpler rivals, we must mark down the complex (flexible) model in estimating the relative divergence from the truth in order to counterbalance its greater propensity for overfitting. It follows that when two or more models fit the data equally well, we should select the simplest one. This implication of the relative-divergence format is important in the discussion of Cartesian skepticism in the next chapter because the BIV hypothesis is designed to fit the data as well as the natural world hypothesis does. We can therefore focus on their complexity.

Some additional remarks are in order. First, to count for K, the parameters must be adjustable. If the value of some parameter is already determined independently of the current investigation and imposed on the model from outside, then there is no room for adjusting the parameter value to accommodate the data. A parameter of this kind does not make the model more flexible, and thus does not make the model more prone to overfitting. So, it does not count for K. Second, probabilistic models to compare by AIC need not be of the same type. In the two examples discussed in this chapter, the simpler of the two models is a special case of the other model: LIN is a special case of QUAD with the parameter $a_2 = 0$, and BNL is a special case of MKV with the parameters $p = q$. When two models to compare are of the same type and one is a special case of the other, it is easy to see which model is simpler (less flexible), but AIC allows us to compare models that are of different types, e.g. a polynomial model and an exponential model. All we need for calculating their AIC scores is their goodness of fit with the data, and the number of adjustable parameters in the model.

It is also noteworthy that the number of variables is not directly related to the flexibility of the model. We already saw some cases where two models have the same number of variables but different numbers of adjustable parameters, e.g. LIN: $y = a_1x + a_0$ and QUAD: $y = a_2x^2 + a_1x + a_0$. There are also cases where adding variables to the model allows us to reduce the number of adjustable parameters. This is particularly interesting when the added variables are unobserved variables (variables whose values are not in the data). Suppose x_1, x_2 and x_3 are independent variables, while y_1 and y_2 are dependent variables. To make it simple, assume that each variable is binary (each variable takes the value 0 or 1). Suppose also that the preliminary investigation indicates that both y_1 and y_2 are dependent on the combinations of x_1, x_2 and x_3 but in different ways. Given these indications, the simplest model we can construct with these variables has eight conditional probabilities for each of the dependent variables y_1 and y_2 since there are $2^3 = 8$ combinations of values for the three binary independent variables x_1, x_2 and x_3. Since y_1 and y_2 are dependent on the combinations in different ways, we need 16 adjustable parameters in the model for $8 \times 2 = 16$ conditional probabilities.

This is all straightforward, but it turns out that we can reduce the number of adjustable parameters by adding an unobserved variable z to the model. Let z be a binary variable that is dependent on the combinations of the values of x_1, x_2 and x_3. We have then eight new adjustable parameters for eight conditional probabilities of z. However, we can reduce the total number of adjustable parameters by making y_1 and y_2 only dependent on z. The values of the variables y_1 and y_2 is still indirectly dependent on the combinations of the values of x_1, x_2 and x_3 because the value of z itself is dependent on them. This makes the model consistent with the preliminary indication that both y_1 and y_2 are dependent on the combinations of the values of x_1, x_2 and x_3. However, the values of x_1, x_2 and x_3 do not have direct impact on y_1 and y_2.[31] Finally, the model makes y_1 and y_2 dependent on z in different ways by distinguishing the conditional probabilities $P(y_1 = 0 \mid z = 0)$ and $P(y_1 = 0 \mid z = 1)$ for y_1, and the conditional probabilities $P(y_2 = 0 \mid z = 0)$ and $P(y_2 = 0 \mid z = 1)$ for y_2. This makes the model consistent with the preliminary indication that y_1 and y_2 are dependent on the combinations of x_1, x_2 and x_3 in different ways. Counting these four conditional probabilities as four adjustable parameters, the new model has a total of $8 + 4 = 12$ adjustable parameters. This makes the new model with the additional variable z simpler than the original model that has 16 adjustable parameters.[32] Cases of this kind reveal how a "realist" model with unobserved variables can be simpler than its "positivist" counterpart with no unobserved variables, and thus its estimated divergence from the truth can be smaller than that of their positivist counterpart even if they fit the data equally well. I will return to this point in the next chapter.

6.6 Models, Theories and Prior Probabilities

This section describes how the relative-divergence format of epistemic evaluation relates to the dual-component format of epistemic evaluation discussed earlier—or more generally, to any Bayesian format of epistemic evaluation, including the Lockean format. As mentioned already, the relative-divergence format does not compete with the dual-component format because the targets of their evaluation are different. The dual-component format evaluates statements that are true or false, while the relative-divergence format evaluates probabilistic models, which are neither true nor false because models only specify general patterns of theories. The two formats complement each other in the following way. First, the relative-divergence format selects a probabilistic model. This is done by letting the competing models generate theories from the data, and estimating their relative divergences from the truth by cross-validation, AIC, etc. Once a probabilistic model M is selected, we let M generate a theory from the data to obtain a probability distribution P, often with the help of observed values for the relevant independent variables. Once a probability distribution P is obtained, then a Bayesian format of epistemic evaluation becomes applicable, viz. we use the Bayesian format to determine whether a statement is acceptable or not on the basis of the probability distribution P.[33]

It is significant that no a priori evaluation is required at any stage of this process. Once we come up with a list of candidate models to evaluate, the relative-divergence format selects a model on the basis of the data. We do need the likelihoods in the process of estimation, i.e. the probabilities that each theory assigns to possible outcomes, but that is just what a probabilistic theory does—i.e. to specify a probability distribution over the partition of outcomes. As long as the candidate models generate probabilistic theories from the data, the relative-divergence format of epistemic evaluation selects a probabilistic model without any a priori evaluation. Once a model is selected, the theory to adopt is the one that the model generates from the data. All we need here are, again, the actual data and the likelihoods of the theories (the probabilities that the theories assign to the data). The adopted theory then assigns probabilities to possible outcomes—often with the help of observed values for the relevant independent variables. We only need the data and the likelihood—and possibly observed values for the relevant independent variables—in the whole process without any a priori evaluation, and yet we obtain a probability distribution over possible outcomes. The relative-divergence format of epistemic evaluation therefore sidesteps the problem of a priori evaluation that troubles Bayesian epistemology.

This may be puzzling to some Bayesians: How can we assign probabilities with no a priori evaluation at all? Given the evidence and the likelihoods, Bayes's Theorem allows us to "update" probabilities. There may be a series of updating because the updated probabilities can be updated further by new evidence. However, a series of updating must start somewhere from some probabilities. The evaluation at that point must be a priori—either based on some a priori principles such as the principle of indifference, or if no such principle is applicable, then by subjective judgment. How can the relative-divergence format of epistemic evaluation do away with a priori evaluation completely?

The answer is a shift in the standard of evaluation from the probability to the divergence from the truth. If we are evaluating theories by their probabilities, then we need prior probabilities somewhere, in addition to the evidence and the likelihoods. However, the relative-divergence format does not evaluate theories by their probabilities. It evaluates them indirectly by evaluating probabilistic models that generate the theories, and the probabilistic models are not evaluated by their probabilities. They are evaluated by the inaccuracy of the model-generated theories (the distance to the truth with regard to the true single outcome, and the divergence from the truth with regard to the true probability distribution). When we assign a probability, and more generally in any quantitative evaluation, our main goal is to make the quantity as close to the true value as possible. It is great if the quantity we assign is exactly the true value, but if that is the goal, we almost always fail. Since the relative-divergence format is a format of quantitative evaluation, its focus is the degree of inaccuracy, and not the probability that the theory is true.

It is also worth noting that a greater discount for a complex model has nothing to do with a priori presumptions about reality. There is, for example, no metaphysical presumption in play that reality is simple, probably simple, etc. Whether reality is simple or complex, a complex probabilistic model is marked down because of its propensity to chase random variation. To reiterate an earlier point, complexity makes a model flexible and flexibility allows the model to chase random variation in the data, which drives the generated theory away from the true probability distribution. This is why a simple model is preferred. Reference to simplicity is confusing at times because it is not immediately clear whether the simplicity at issue is a feature of the model or a feature of reality. In the relative-divergence format of epistemic evaluation it is simplicity of the model that counts, and this has nothing to do with the aesthetic merits of simplicity. The crucial feature is flexibility, i.e. a simple model is preferred because simplicity limits the flexibility of the model.

Some Bayesians may feel that my account of how the relative-divergence format does away with a priori evaluation misconstrues the role of prior probabilities in Bayesian epistemology. Prior probabilities should not be considered probabilities assigned in the absence of any relevant empirical evidence, it may be suggested, because we have no idea what such probabilities are like. We always have some prior experience relevant to the subject of our interest. So, prior probabilities are probabilities assigned prior to the present investigation, which are still based on our relevant experience from the past. Those people may worry that the relative-divergence format, which does not take prior probabilities into account, leaves no room for relevant prior experience to influence the evaluation. For example, suppose BNL performed much better than MKV in the past for many slot machines, though the parameter values were different for different machines. We have then good reason to think BNL will perform better than MKV for the next machine even if we have no data on it yet, and this prior preference for BNL may persist even if the initial data on the next machine favors MKV. In general, it should count for something if one model performed better than another in similar cases in the past. Bayesianism can account for prior experience of this kind in the form of prior probabilities. Some Bayesians may find it troubling that prior probabilities play no role in the relative-divergence format.

There are, of course, certain ways the prior experience influences model selection. First, since we cannot investigate all the probabilistic models that we can think of, models that performed poorly in similar cases in the past may not be included in the list of candidates for the new case. If a model is not in the list of candidates, then obviously it won't be selected. This is not inconsistent with the relative-divergence format of epistemic evaluation, or any format of epistemic evaluation for model selection, because the decision on which models to evaluate is not part of the evaluation.[34] The prior experience plays no part in the evaluation of models that are included in the list.[35] This may seem like a distinction without a difference, but there is actually an important difference. Suppose someone mentions MKV as a possible model in response to your selection of BNL. It is one thing to say that you did not evaluate MKV because your resource is limited and MKV performed poorly in similar cases. Anyone who disagrees with you can then take up the task for herself. It is absurd, on the other hand, to say that you evaluated MKV and it performed well, but you rejected it anyways because it performed poorly in similar cases in the past.

If a model is in the list of candidates to evaluate, then only the data can justify its rejection. This is in contrast with Bayesianism that can cite a low prior probability for rejection. It is still possible to use the relative-divergence format to reject MKV for the reason of its poor track record in similar cases, but that requires an extended model that covers not just the new slot machine but all slot machines of the same type. The extended model would have additional variables not present in BNL or MKV whose values would determine the probability profiles of different slot machines. If unification by an extended model fails and we are left with a separate model for each slot machine, then we should not allow prior experience with other slot machines to influence model selection in the present case.

Of course, not all experience is systematically organized in the form of a data set. We often rely on our intuitive judgments grounded in the past experience even though they are fuzzy and informal. This also seems true of model selection. For example, if we do not have systematic data on the previously encountered slot machines of the same type, we cannot apply the relative-divergence format to an extended model that covers these machines. Some may want to argue it is still reasonable to dismiss MKV by intuitive judgments grounded in the past experience, especially if people familiar with those machines make the same intuitive judgments. As mentioned already, it is consistent with any format of epistemic evaluation not to include a model in the list of candidates based on intuitive judgments. It is unreasonable, however, to reject a well-performing model on the list by fuzzy and informal intuition.[36] If the available data does not justify its rejection, those who question the model must keep collecting data, perhaps extensive data on other slot machines of the same type, until the total evidence justifies the rejection.

This chapter introduced the relative-divergence format of quantitative epistemic evaluation to complement the dual-component format of qualitative epistemic evaluation that makes use of probabilities assigned by quantitative epistemic evaluation. The chapter examined various features of the new format, but the most important among them is the role of simplicity of the model in the estimation of the relative divergence from the truth. When competing models generate theories that fit the data equally well, we should choose the simplest among them. With this in mind, I now return to the challenge of Cartesian skepticism.

Notes

1 The notion of "overfitting" in particular has drawn the attention of many working scientists lately. Philosophers of science are also exposed to the Akaike Information Criterion (AIC) mainly through the work of Malcolm Forster and Elliott Sober (Forster & Sober 1994).

2 <−2, 2> abbreviates $x = -2$ and $y = 2$, and <1, 5> abbreviates $x = 1$ and $y = 5$.

3 It is sometimes stipulated that QUAD is correct only if $a_2 \neq 0$, to make LIN and QUAD mutually exclusive. I make no such stipulation.

4 "G" in P(G) abbreviates the statement that the displayed color is green. There is no need to mention the probability of B here because $P(\text{G}) = p$ entails $P(\text{B}) = 1 - p$ given the assumptions of the case.

5 ML is more general than the method of mean squared deviation (MSD), also known as mean squared error (MSE). MSD chooses the parameter value that minimizes the mean squared deviation of the data from the predicted (by the parameter value) average outcome. Since MSD is limited to cases where the outcome is quantitative, it is not appropriate in the present case where the outcome is qualitative (G or B).

6 Given BNL and any parameter value p, we can calculate the probability distribution (the binomial distribution) over the numbers of Gs among n outcomes: Given $P(\text{G}) = p$ for each outcome, the probability that exactly k out of n outcomes are green is $C(n, k)p^k(1 - p)^{n-k}$, where $C(n, k)$ is the binomial coefficient $n!/[k!(n - k)!]$.

7 Given BNL and the parameter value $p = 0.5$, the probability of obtaining the exact sequence DS (not just the ratio of Gs) is 0.5 to the power of 10, which is about 0.00098. Any other parameter value makes the probability lower.

8 When a sequence is a Markov chain, the probability of each outcome in the sequence only depends on the immediately preceding outcome. As we will see shortly, we can regard a Bernoulli distribution as a special case of a Markov chain.

9 Given MKV and the parameter values $p = 0.4$ and $q = 0.6$, the probability of obtaining the exact sequence [G]GBGBBGBGBB is $0.4 \times 0.6 \times 0.6 \times 0.6 \times 0.4 \times 0.4 \times 0.6 \times 0.4 \times 0.6 \times 0.6 = 0.4^4 \times 0.6^6$, which is about 0.0012. Any other combination of parameter values makes the probability lower.

10 As mentioned in Chapter 4, the idea of evaluating theories by their distance to the truth goes back to Popper (1963; 1972). Popper and his supporters try to answer the question of "verisimilitude" (truthlikeness) by comparing the true and false contents of theories. In this chapter the divergence from the truth is understood, instead, by the inaccuracy of a probability distribution.

11 The probabilistic versions contain an "error term" to account for random variation. For example, LIN becomes $y = a_1x + a_0$ $y + \varepsilon$ and QUAD becomes $y = a_2x^2 + a_1x + a_0 + \varepsilon$.

12 See, for example, Leitgeb and Pettigrew (2010) and Pettigrew (2016).

13 We encountered a similar situation in Chapter 4, viz. the amount of information the hypothesis h carries is inversely related to the probability assigned to h, but there are many inverse functions of the probability to choose from. As we will see shortly, the appropriate function in both cases is $-\log P(h)$. See Roche and

Shogenji (forthcoming) for an analysis of the relation between information and inaccuracy.

14 Sometimes the term "the expected value" is used instead of "the average value", but the term "expected" is misleading in this case because we cannot form an expectation without knowing the true distribution *T*. When we use *P* as our best estimate, we actually expect the post hoc inaccuracy to be $P(G)SR(P; G) + P(B)SR(P; B)$ instead of $T(G)SR(P; G) + T(B)SR(P; B)$. The average post hoc inaccuracy is not the post hoc inaccuracy we actually expect, but the post hoc inaccuracy we *should* expect.

15 The subscripts in "SR_C", "SR_{SC}", "SR_R" and "SR_L" stand for "Complement", "Squared Complement", "Reciprocal" and "Logarithm", respectively. For more on the candidates of a scoring rule, see Murphy and Winkler (1984), Winkler (1967, 1969, 1971, 1994), and Winkler and Murphy (1968).

16 Though Strict Propriety is embraced across the board, it is usually motivated in a different way, where *T* is the forecaster's best estimate (instead of the true distribution) while *P* is the estimate that the forecaster publicly announces. Strict Propriety discourages the forecaster to dishonestly announce *P* that is different from *T* because if the scoring rule is strictly proper, the forecaster can minimize the expected inaccuracy only by publicly announcing *T* itself.

17 Those who are familiar with the literature would have noticed that "global scoring rules" are ignored in my discussion. The reason is that a global scoring rule is not a monotonic function (and thus not an inverse function) of the probability assigned to the true statement when the partition has three or more statements. See Roche and Shogenji (forthcoming) for more on this point.

18 In the special case where *P* is identical to *T*, the cross entropy is the Shannon entropy H(*T*):

$$H(T) = -\sum_{i=1}^{n} T(y_i) \log T(y_i)$$

The Kullback-Leibler divergence is also known as the relative entropy, and can be obtained by subtracting the Shannon entropy from the cross entropy: $KL(T \parallel P) = H(T, P) - H(T, T)$. When understood in this way, $KL(T \parallel P)$ measures an increase in entropy, or loss of information, for accepting *P* instead of the true probability distribution *T*.

19 I leave the base *b* of logarithm unspecified because the answer does not depend on it. As long as *b* is greater than one, $\log_b x$ is an increasing function of *x*. I will use natural logarithm (i.e. with Euler's number *e* as the base) later in the chapter for technical reasons.

20 We can calculate the reliability of the selected model from the data size, variance, etc.

21 See Kohavi (1995) and Arlot and Celisse (2010) among others for discussions of appropriate versions of cross-validation for different applications.

22 See Emery (2015) for related discussions about the chance (the objective probability).

23 Kuipers (2000, Ch. 7) introduces a similar distinction between the "actual" truth and the "nomic" truth, and argues that truthlikeness should be measured relative to the nomic truth. He articulates the concept of the nomic truth in modal terms (by possibility and impossibility).

24 Some people may still worry that even the best global model may pick up accidental relations in the sense that the relations hold globally but not counterfactually. It is unclear whether this is a problem, or even how it should be understood. It seems that the best global model for the world we live in is good enough. If there is some reason to take the counterfactual worry seriously, we may (conceptually) incorporate "counterfactual data" into the full data for the purpose of defining the full truth.

25 "a" in $P(G \mid a)$ and $Q(G \mid a)$ abbreviates $x = a$.

26 See Burnham and Anderson (2002, Ch. 4) for technical details of this approach.

27 The situation is different if a third model M_3 is also under consideration, where M_3 is the weighted average $\omega[ML(M_1, D)](G \mid \alpha) + (1 - \omega)[ML(M_2, D)](G \mid \alpha)$ of the models M_1 and M_2, and the value of the parameter ω is determined from D by the method of maximum likelihood. If the list of candidates consists of M_1, M_2 and M_3, we need to compare the three models by the relative-divergence format of epistemic evaluation. Note, however, that M_3 is a lot more flexible and thus prone to chase random variation. It contains all adjustable parameters of M_1, all adjustable parameters of M_2, and then the additional parameter ω, which is also adjustable.

28 Given this data sequence, the parameter values for MKV are $p = 0$ and $q = 1$, and there is no training error at all—the generated theory fits the data perfectly. The cross-validation score is also zero. Meanwhile, the parameter value, the training error and the cross-validation score for BNL on this data all remain the same as on the data sequence DS mentioned earlier.

29 For example, AIC is a good approximation when the data size is large, and the model must be "regular". AIC_c (Hurvich and Tsai 1989) is a good approximation even when the data size is small, and WAIC (Watanabe 2009) is applicable even if the model is "singular".

30 So, we can choose from two families of methods in applying the relative-divergence format of epistemic evaluation. Though it has been shown that AIC and leave-one-out cross-validation are asymptotically equivalent (Stone 1977), there can be disagreement in their estimates when the models to compare are almost as good as each other.

31 In other words, the variable z "screens off" any combination C of the values of x_1, x_2 and x_3 from y_1 and y_2 in the sense that $P(y_1 = 0 \mid z = 0 \wedge C) = P(y_1 = 0 \mid z = 0)$ and $P(y_1 = 0 \mid z = 1 \wedge C) = P(y_1 = 0 \mid z = 1)$, and similarly $P(y_2 = 0 \mid z = 0 \wedge C) = P(y_2 = 0 \mid z = 0)$ and $P(y_2 = 0 \mid z = 1 \wedge C) = P(y_2 = 0 \mid z = 1)$.

32 Obviously, the values of the unobserved variables are not in the data, but the model can still generate a theory because the values of the unobserved variables can be determined by the values of independent variables and the combination of (possible) parameter values. We select the combination of parameter values that make the values of the dependent variables y_1 and y_2 closest to their actual values in the data.

33 In the dual-component format we use P to obtain two probabilities $P(h \mid b)$ and $P(h \mid e \wedge b)$, both of which are needed for determining the degree of epistemic worthiness $J(h, e \mid b)$.

34 In terms of the distinction drawn in Chapter 1 between the context of discovery and the context of evaluation (Reichenbach 1938), the decision on which hypotheses to investigate belongs to the context of discovery, and not the context of justification where the hypotheses are evaluated.

35 The point only applies to the selection of a probabilistic model by the relative-divergence format. Once a probabilistic model is selected, we need to distinguish between $P(h \mid b)$ and $P(h \mid e \wedge b)$ to determine the degree of epistemic worthiness $J(h, e \mid b)$ in the dual-component format, where $P(h \mid b)$ can be considered the prior probability, though it is used to determine the informativeness of h in the dual-component format.

36 Note that intuitive judgments can be part of the relative-divergence format of epistemic evaluation if they are amenable to statistical analysis. For example, we may collect intuitive judgments made by the experts, and construct a model in which those judgments are a variable in the model. If the model performs well in cross-validation, the relative-divergence format of epistemic evaluation may select the model that incorporates the expert judgements.

7 Cartesian Skepticism Defeated

7.1 Solution to the Problem of Equal Likelihood

This final chapter applies the relative-divergence format of epistemic evaluation to solve the problem of equal likelihood, and then to solve Cartesian skepticism. Let us recall how the problem of equal likelihood arises to challenge the practice of induction. When some hypothesis h is supported by the actual body of evidence e_A obtained so far, we rely on h to predict an occurrence of some event o. For example, we predict that the sun will rise in the east tomorrow morning because the past observations support the hypothesis that the sun rises in the east in the morning. However, when it seems reasonable to predict an occurrence of o based on some hypothesis h, the skeptic constructs an alternative hypothesis h^* that has the same likelihood as h does in the sense that $P(e_A \mid h \wedge b) = P(e_A \mid h^* \wedge b)$, but is inconsistent with o. For example, that the sun rises in the east in the morning till today but rises in the *west* in the morning thereafter. This hypothesis entails all the past observations of the sunrise, just as the original hypothesis does, but it predicts that the sun will rise in the west tomorrow morning.

Customary responses to inductive skepticism try to establish that the standard hypothesis h is superior to the alternative hypothesis h^*, but that is not enough to defend our inductive practice. The skeptic argues for the third option of only accepting their disjunction $h \vee h^*$ without committing to either disjunct. The issue of inductive skepticism is therefore not resolved by the comparison of h and h^*. We need to compare h and $h \vee h^*$, and the dual-component format of epistemic evaluation reveals that $h \vee h^*$ is epistemically more worthy than h when h and h^* are equally likely in the sense that $P(e_A \mid h \wedge b) = P(e_A \mid h^* \wedge b)$. This means that we should remain non-committal between h and h^*, and thus should not predict that the sun will rise in the east tomorrow morning based on h because the prediction is inconsistent with the other disjunct h^*. The same reasoning makes any prediction based on any hypothesis unreasonable no matter how well the hypothesis is supported by the evidence obtained so far. This is the way the problem of equal likelihood threatens the practice of induction.

As noted in Chapter 5, the problem of equal likelihood is orthogonal to the new riddle of induction (Goodman 1955), where the challenge arises in the form of an alternative predicate, such as "grue", grounded in an alternative conceptual space. Even if we solve the new riddle of induction by successfully defending our current conceptual space in which grue is not a natural property, the problem of equal likelihood persists since the choice of a conceptual space plays no part in it. I set aside the new riddle of induction and focus on the problem of equal likelihood, assuming that our current conceptual space is defensible.

I now present a solution to the problem of equal likelihood by the relative-divergence format of epistemic evaluation. The strategy is to rule out the twisted alternative theory h^* before applying the dual-component format, on the ground that no probabilistic model generates h^* from the actual body of evidence obtained so far. I will illustrate the point by an example.[1] Recall the unconventional slot machine introduced in the previous chapter. When you insert a dollar, the machine displays its default color of green (G). Then, each time you press the button, it displays either green (G) or blue (B). I will use a lopsided data sequence so that the standard inductive reasoning is compelling. Let DL be the sequence [G]GGGBGGGGGG, of which the first [G] is the default color that does not count. As before, an obvious model to consider is BNL (the Bernoulli distribution) and BNL generates the theory h below from the sequence DL.

$$\text{BNL: } P(\text{G}) = p$$
$$h\text{: } P(\text{G}) = 0.9$$

Based on the theory h, we can predict that the next outcome will be a G with the probability of 0.9. This is a familiar kind of inductive reasoning.

The skeptic constructs an alternative theory h^* by following the recipe above, viz. h^* states that h holds till time t but f holds thereafter, where f is inconsistent with the prediction of G for the next outcome.

$$h\text{: } P(\text{G}) = 0.9$$
$$f\text{: } P(\text{G}) = 0$$

The skeptic combines h and f to construct the alternative theory h^*. Since the outcomes in the case constitute a discrete sequence, we make the position in the data sequence an independent variable x, instead of referring to time. For example, $P(\text{G}x)$ is the probability that the xth outcome is a G. The resulting theory h^* is as follows.

$$h^*\text{: } h \text{ [i.e. } P(\text{G}x) = 0.9] \textit{ for } x \leq 10$$
$$f \text{ [i.e. } P(\text{G}x) = 0] \textit{ for } x > 10$$

Because of the way h^* is constructed, h and h^* are equally likely up to $x = 10$. So, given the data sequence up to $x = 10$, the dual-component format states that the disjunction $h \vee h^*$ has a greater epistemic worth than either disjunct, i.e. $J(h, \text{DL} \mid b) < J(h \vee h^*, \text{DL} \mid b)$ and $J(h^*, \text{DL} \mid b) < J(h \vee h^*, \text{DL} \mid b)$. We should therefore remain non-committal between h and h^*, according to the skeptic, and this makes it unreasonable to assign a high probability to G for the next outcome.

The solution is to rule out h^* before we apply the measure J to the competing theories. The ground for elimination is the absence of a probabilistic model that generates h^* from the actual body of evidence DL: [G]GGGB-GGGGGG. There are, of course, many probabilistic models that generates h^* from a suitable body of evidence, but no probabilistic model generates h^* from DL. Take, for example, the model BNL* with a change in the probabilities beyond $x = k$.

$$\text{BNL}^*\!: P(\text{G}x) = p \ \textit{for} \ x \le k$$
$$P(\text{G}x) = q \ \textit{for} \ x > k$$

BNL* would generate h^* from some data sequence. Consider, for example, the following sequence of 20 data points: [G]GGGBGGGGGG-BBBBBBBBBB. Given this longer data sequence, the best parameter values are clearly $k = 10$, $p = 0.9$ and $q = 1$ because nine out of the first ten outcomes are Gs while all outcomes thereafter are Bs. So, BNL* generates h^* from some data sequence, but not from the actual data DL in the present case: [G]GGGBGGGGGG. Since we only have the first ten outcomes, there is simply no way of determining the probability q that the outcome will be a G after the first 10 outcomes.

It may be suggested that f need not be inconsistent with the prediction of G for the next outcome. The skeptic only needs to make the probability of G low. For example, the skeptic may reverse the probabilities of G and B after $x = 10$ so that the alternative theory $h\dagger$ assigns a low probability to G after $x = 10$.

$$h\!: \ P(\text{G}) = 0.9$$
$$f\dagger\!: P(\text{G}) = 0.1$$
$$h\dagger\!: h \left[\text{i.e. } P(\text{G}x) = 0.9\right] \ \textit{for} \ x \le 10$$
$$f\dagger \left[\text{i.e. } P(\text{G}x) = 0.1\right] \ \textit{for} \ 10 < x$$

An additional advantage of reversing the probabilities is that the skeptic does not need two adjustable parameters p and q for the probability of G because $q = 1 - p$ is dependent on p. This allows a simpler model to generate $h\dagger$.

$$\text{BNL}\dagger\!: P(\text{G}x) = p \ \textit{for} \ x \le k$$
$$P(\text{G}x) = 1 - p \ \textit{for} \ x > k$$

Again, BNL† would generate $h\dagger$ from some data sequence, for example, the sequence of 20 data points: [G]GGGBGGGGGGBBBBBBBGBB. However, BNL† still fails to generate $h\dagger$ from the actual data DL in the present case: [G]GGGBGGGGGG. There is no way of determining from the first ten data points when the reversal occurs—after 10 outcomes, after 11 outcomes, after 12 outcomes, etc. When the reversal occurs is crucial for prediction: $P(\text{G}_{11})$ is low if $k = 10$ but $P(\text{G}_{11})$ is high if $k \ge 11$.

To summarize, it is conceded that whenever a reasonable theory is proposed, the skeptic can use a general recipe to construct an alternative theory that has the same likelihood as the proposed theory does, but whose prediction contradicts that of the proposed theory. However, the alternative theory is so twisted that no probabilistic model generates it from the given data. In general, a probabilistic model is too complex (too flexible) for the evidence if it does not generate a theory from the data. If no model generates the theory from the given data, we cannot even estimate its relative divergence from the truth because the estimate depends on the complexity (flexibility) of the model that generates the theory. We must therefore rule out the twisted theory as inappropriate for the given data.

7.2 Failure of Two Maneuvers

The previous section pointed out that we need not consider the skeptic's twisted alternative theory because no probabilistic model generates it from the actual data. There are, however, some maneuvers the skeptic may use to avoid this difficulty. This section considers two unusual ways of letting a probabilistic model generate a twisted alternative theory from the actual data. One of them brings in non-adjustable parameters, while the other makes use of expedient variables.

Recall, first, that it is the number of *adjustable* parameters in the model that counts for its degree of complexity (flexibility). This is because non-adjustable parameters do not make the model flexible. With no increase in flexibility, non-adjustable parameters do not make the model more prone to chase random variation. For example, the model $y = a_0 + a_1x_1 + 0.38x_2$ with two independent variables, x_1 and x_2, is no more flexible than the model $y = a_0 + a_1x_1$ with only one independent variable because both of them have exactly two adjustable parameters, a_0 and a_1. The skeptic may exploit this point to argue that a key parameter in the model is non-adjustable, so that the model with limited flexibility generates a twisted alternative theory from the actual data.

Let us return to the example of the model BNL generating the theory *h* from the lopsided data sequence DL.

BNL: $P(\text{G}) = p$

DL: [G]GGG*B*GGGGGG

h: $P(\text{G}) = 0.9$

The theory *h* predicts that the next outcome is a G with the probability of 0.9, but the skeptic introduced the following alternative.

$$\text{BNL}\dagger: \; P(\text{G}x) = p \;\; \textit{for} \;\; x \leq k$$
$$P(\text{G}x) = 1 - p \;\; \textit{for} \;\; x > k$$
$$h\dagger: \; P(\text{G}x) = 0.9 \;\; \textit{for} \;\; x \leq 10$$
$$P(\text{G}x) = 0.1 \;\; \textit{for} \;\; x > 10$$

The alternative theory $h\dagger$ is designed to have the same likelihood as h does up to $x = 10$, but $h\dagger$ assigns a low probability $P(G_{11}) = 0.1$ to G for the next outcome. In response to this challenge, it was pointed out there that BNL† does not generate $h\dagger$ from the data sequence DL because given DL it is not just $k = 10$ that maximizes the likelihood. Any value $k \geq 10$ maximizes it, so BNL† does not generate $h\dagger$ which has the specific value $k = 10$.

Here is a further twist. With the use of non-adjustable parameters, the skeptic can overcome the problem of underdetermination of k by the following model $\text{BNL}\dagger_{10}$ in which $k = 10$ is a non-adjustable parameter.

$$\text{BNL}\dagger_{10}: P(Gx) = p \text{ } for \text{ } x \leq k = 10$$
$$P(Gx) = 1 - p \text{ } for \text{ } x > k = 10$$

The model $\text{BNL}\dagger_{10}$ generates $h\dagger$ from the actual data DL because there is no possibility of $k = 11, 12, \ldots$ any more. Of course, this is only a charade. There is no reason to set $k = 10$ other than the wish to generate the twisted theory $h\dagger$. Perhaps in an effort to make the idea look more palatable, the skeptic may include all the models $\text{BNL}\dagger_1$, $\text{BNL}\dagger_2$, ... (with $k = 1, 2, \ldots$, respectively) in the list of candidates, to maintain that all these models are treated equally. However, the inclusion of models with different non-adjustable parameter values brings back the problem of underdetermination: Given the actual data DL, the model $\text{BNL}\dagger_{10}$ is no better than the models $\text{BNL}\dagger_{11}$, $\text{BNL}\dagger_{12}$, ... in the list because they fit the data equally well and are equally simple. Even though the model $\text{BNL}\dagger_{10}$ generates $h\dagger$ from DL, we are unable to select the model $\text{BNL}\dagger_{10}$ from the list on the basis of the actual data DL.

So, the skeptic's declaration of $k = 10$ as a non-adjustable parameter is arbitrary, and it should be blocked by some principle on the adjustability of parameters. Some such principle is needed anyways because without some principle on the adjustability of parameters, we would be unable to guard against overfitting in model selection. For example, we may formulate a theory with numerous parameters and choose the parameter values in light of the data, and then declare post hoc that all the parameter values are non-adjustable to make the model the simplest possible with no adjustable parameters. That would destroy the distinction between models and theories that is crucial to guarding against overfitting. Without some principle, any theory would be a model in its own right with no difference in complexity (flexibility), so that only the goodness of fit would count.

There is a simple principle that serves the purpose: All parameters in a probabilistic model are adjustable by default unless they are shown to be otherwise. Since the parameter value is a quantity, and since any particular quantity must be supported in some way, the parameter is adjustable by default. There are, of course, many cases where some parameters of a model are genuinely non-adjustable. For example, we cannot adjust the Boltzmann constant in a climate model to fit the data better. Its value $k = 1.38064852 \times 10^{-23}$ (J/K) is established independently of the climate data. I will call a parameter "exogenous" if its value is established independently of the present research. A parameter that is not exogenous is "endogenous" and its value must be determined by the present research. To restate the principle, all parameters in a probabilistic model are endogenous unless they are shown to be otherwise. This principle rules out models with arbitrary parameter values, such as $\text{BNL}\dagger_{10}$.

Declaring a key parameter non-adjustable is not the only way of letting a probabilistic model generate a twisted alternative theory from the data. The skeptic can achieve this by introducing an *expedient variable*. I will illustrate the maneuver with the same example of BNL generating *h* from DL. To challenge *h*, the skeptic needs some model that generates a twisted theory from DL, which predicts the next outcome is a B instead of a G. The prediction would then extend the sequence in the following way: [G] GGGBGGGGGG(B), where the last B in parentheses is predicted by the twisted alternative theory. The skeptic therefore wants a theory to say that the 4th and 11th outcomes are Bs. To construct a model that generates such a theory, the skeptic may search for some variable *x* that is already known to take a certain value exactly in the 4th and 11th occurrences. Suppose such a variable *z* is found. To simplify the case, let *z* be a binary variable, and $z = 1$ in the 4th and 11th occurrences while $z = 0$ in all other occurrences in the extended sequence. With the expedient variable *z* in hand, the skeptic now proposes the expedient-variable model M_{EV} below with a single (adjustable) parameter *p*, which generates the theory h_{EV} with $p = 1$ from DL.

$$M_{\text{EV}}: P(\text{G}) = |z - p|$$
$$h_{\text{EV}}: P(\text{G}) = |z - 1|$$

The theory h_{EV} fits DL perfectly: It assigns the probability zero to the 4th outcome being a G, and probability one to each of the nine other outcomes in DL being a G. This means that h_{EV} fits the data even better than *h* does, and the model M_{EV} is as simple as the model BNL with only one adjustable parameter *p*. So, M_{EV} is superior to BNL in the case, but the theory h_{EV} generated by M_{EV} from DL predicts that the next outcome is not a G.

Of course, any independent variable of the model must have its values available in the data, and there is no guarantee that an expedient variable of this kind can be found. However, the skeptic may take the shotgun approach to data collection and assemble "big data" to increase the number of available variables dramatically. An extensive search will eventually turn up some variable with the desired feature. We will not be able to block the maneuver for the reason of availability alone. Fortunately, we already have a different reason for blocking the maneuver. Recall the discussion of the full global truth as our aim in the previous chapter. Given this aim, a local model on a particular subject matter is rejected if the theory it generates does not fit well with the best theories on related subject matters. More generally, a local theory must be consistent with the general suppositions about the structure of reality that are supported by other local theories. Since an expedient variable is included in the model for the sole purpose of generating the twisted theory, and has no relevance to the subject matter, the resulting theory cannot be integrated into global theories. For example, the skeptic may find out that the SP 500 Index went up in all and only those years when it was sunny in Kathmandu on the Groundhog Day. No matter how good the fit is, and no matter how simple the model is, it is unacceptable because it is inconsistent with our general suppositions about the economy and the climate that are supported by many local theories.

There may be a protest that this attitude will stifle radical innovations that overturn the current general suppositions. However, the principle does not have such an implication. It is perfectly fine to take issue with the current general suppositions. There are many cases in the history of science where a hypothesis that is inconsistent with the general suppositions of the time was proposed in order to overcome some difficulties, and some of them eventually became an orthodoxy in the field. I am not ruling out such a possibility. What is unacceptable is an ad hoc suspension of the general suppositions just for one instance of local model selection. If we introduce a theory that is inconsistent with the current general suppositions about the structure of reality, we need to search for an alternative set of general suppositions and test them in cases where revising the general suppositions affects local theories.

7.3 Cartesian Skepticism and the BIV Models

The remainder of the chapter addresses Cartesian skepticism. To recall, the skeptic introduces the hypothesis that we are brains in vats (BIVs) kept alive and wired to a supercomputer that supplies us with the kind of sensory input brains in skulls would receive from the natural environment. In this scenario what we take to be the natural world is only a virtual world generated by the supercomputer, but we cannot tell the difference. Unlike skepticism about induction, the stakes are not high in this challenge because even if we are BIVs, there will be no surprising turn of events at any point as far as our sensory experience is concerned. So, if we cannot solve Cartesian skepticism, we have the option of settling for the modest goal of epistemic justification up to empirical equivalence with no change in the expected sensory experience. As it turns out, however, we can solve Cartesian skepticism by the relative-divergence format of epistemic evaluation.

The first thing to note is that the BIV hypothesis is a highly abstract model with little specificity. The BIV hypothesis itself does not help us predict what sensory input the BIVs will receive. It also lacks the details of the causal process by which our sensory experience is produced. These are not immediate threats to the BIV hypothesis. The skeptic can concede that the BIV hypothesis is a highly abstract model, but point out that the natural world hypothesis is also a highly abstract model, whose layers of submodels—each with variables and parameters—are identified and filled out as we obtain more sensory experience. In the same way the layers of submodels of the BIV hypothesis—each with variables and parameters—are also filled out as we obtain more sensory experience, the skeptics can maintain. So, the lack of specificity is not a unique problem of the BIV hypothesis, but it gives rise to some serious questions. I will discuss the specificity of the contents of our sensory experience first (Sections 3–5), and then the specificity of the causal process responsible for our sensory experience (Sections 6–7). It is the latter that leads to the demise of Cartesian skepticism.

It is convenient to have names for the two components of the BIV hypothesis. I will call part of the BIV hypothesis that describes the contents of our sensory experience "$h_{BIV/CON}$" while calling part of the BIV hypothesis that describes the causal process that produces our sensory experience "$h_{BIV/PRO}$". To begin with the former, the lack of specificity in $h_{BIV/CON}$ prompted some epistemologists to distinguish two versions of $h_{BIV/CON}$.[2] The *generic* version $h_{BIV/CON-G}$ only states that the supercomputer is programed to supply the BIVs with sensory input, while the *specific* version $h_{BIV/CON-S}$ states more specifically that the supercomputer is programmed to supply the BIVs with sensory input so as to deceive us into thinking that we live in a natural world. The distinction lays the ground for the following Bayesian argument against Cartesian skepticism (BonJour 1985, Sec. 8.4; Huemer 2016). If the skeptic proposes the generic version $h_{BIV/CON-G}$, it is extremely unlikely that we (BIVs) have the kind of sensory experience e_A we actually have, i.e. the kind we would have in the natural environment. The two hypotheses $h_{BIV/CON-G}$ and h_{NW} are therefore not empirically equivalent, and we can resolve the dispute between them empirically, i.e. from the low likelihood $P(e_A \mid h_{BIV/CON-G})$ vis-à-vis $P(e_A \mid h_{NW})$ it follows by Bayes's Theorem that the sensory experience e_A disconfirms (lowers the probability of) the generic hypothesis $h_{BIV/CON-G}$.

Meanwhile, if the skeptic proposes the specific version $h_{BIV/CON-S}$, as is customary, then the BIVs have the kind of sensory experience the brains in skulls have in the natural environment. This makes the two hypotheses $h_{BIV/CON-S}$ and h_{NW} empirically equivalent. However, the prior probability of $h_{BIV/CON-S}$ is then extremely low because the specific kind of sensory experience is one of numerous kinds of sensory experience the supercomputer can be programed to generate in the BIVs. Due to its extremely low prior probability, the conditional probability $P(h_{BIV/CON-S} \mid e_A)$ of the specific version of the BIV hypothesis is much lower than that of the natural world hypothesis $P(h_{NW} \mid e_A)$ even if the evidence e_A is neutral between them (even if the evidence keeps the proportion of their probabilities unchanged). The upshot of the argument is that, regardless of the skeptic's choice between $h_{BIV/CON-G}$ and $h_{BIV/CON-S}$, the natural world hypothesis h_{NW} is more probable than is the BIV hypothesis, given our actual sensory experience.

The argument is not convincing as it stands because it relies on the Lockean format of epistemic evaluation that focuses on the conditional probability of the hypothesis given the evidence. As shown in Chapter 4, there is good reason to abandon the Lockean format in favor of the dual-component format, and skepticism is vindicated by the dual-component format, viz. when two hypotheses are empirically equivalent, as in the case of $h_{BIV/CON\text{-}S}$ and h_{NW}, the dual-component format recommends that we remain neutral between them regardless of their prior probabilities. However, the argument above points to a line of reasoning in support of the natural world *model* by the relative-divergence format of epistemic evaluation.

It is noted already that the details of the natural world hypothesis and the BIV hypothesis are filled out as we obtain sensory experience. This means that these hypotheses are actually models that generate theories from the body of evidence. Further, they are probabilistic models because theories they generate make probabilistic predictions of sensory experience. To underscore this point, I will use "M_{NW}" and "M_{BIV}" to refer to the natural world model and the BIV model, respectively. Seeing the BIV hypothesis as a model makes it easier to understand the relation between the two versions of the BIV hypothesis, viz. the generic version $M_{BIV/CON\text{-}G}$ is the root model while the specific version $M_{BIV/CON\text{-}S}$ is its submodel. Prior to obtaining any evidence, the appropriate model to start with is the root model $M_{BIV/CON\text{-}G}$ since there is no reason to restrict it to the submodel $M_{BIV/CON\text{-}S}$ in the absence of empirical input.

With this in mind, let us take another look at the first horn of the dilemma in the (unsuccessful) Lockean argument against Cartesian skepticism. The key point there is that the generic version of the BIV hypothesis has a lower likelihood $P(e_A \mid h_{BIV/CON\text{-}G})$ with regard to our actual sensory experience than does the natural world hypothesis, $P(e_A \mid h_{NW})$, and the reason is that the hypothesis $h_{BIV/CON\text{-}G}$ is compatible with a wider range of sensory experience than is the natural-world hypothesis h_{NW}. To put this in the language of model selection, the model $M_{BIV/CON\text{-}G}$ is more flexible in accommodating possible sensory experience than is the model M_{NW}. As explained in the previous chapter, the relative-divergence format of epistemic evaluation favors the less flexible model over the more flexible model when two models fit the evidence equally well. How well do the two models $h_{BIV/CON\text{-}G}$ and h_{NW} fit the evidence, then? If the supercomputer is programed to deceive BIVs into thinking that they live in a natural world, then the model $M_{BIV/CON\text{-}G}$ eventually generates the same theory of the world from our actual sensory experience as M_{NW} does, so that the two models fit the evidence equally well. We should then select the less flexible model M_{NW} over the more flexible model $M_{BIV/CON\text{-}G}$ by the relative-divergence format, the argument concludes.

I just converted an argument from different likelihoods into an argument from different flexibilities, where the former supports a theory while the latter supports a model. The conversion of this kind is helpful beyond this particular instance because in many Bayesian arguments from different likelihoods, the competing hypotheses are actually probabilistic models. Recall, for example, the argument for the representational parsing of the visual input in Chapter 2. Given the 2-D visual input, we construct a 3-D model of the natural world by locating our visual sensor at the center of the 3-D world. Our 2.5-D visual experience (of surfaces with varying depths) is then taken to be a representation of the facing side of the nearest non-transparent object in the direction of the sensor. In Bayesian terms the reason for accepting this hypothesis is different likelihoods: The likelihood of the representational hypothesis $P(e_A \mid h_R)$ is much higher than the likelihood of the non-representational hypothesis $P(e_A \mid h_{NR})$, where e_A is the actual visual experience. However, the hypotheses h_R and h_{NR} are models whose details are filled out by the actual sensory experience. If we put the comparison in terms of model selection, the model M_{NR} is much more flexible in accommodating possible sensory experiences than is the model M_R, and that is why we should choose the latter if the two models accommodate the actual evidence equally well.

Some people may be puzzled by the claim that the representational model M_R is less flexible than the non-representational model M_{NR}. Part of the representational model is the 3-D model of the world, for which we must add unobserved variables (variables whose values are not in our sensory experience) to the model. It may seem that containing unobserved variables in addition to observed variables makes the model more complex than the non-representational model that does not need unobserved variables. However, as shown in Chapter 6, the complexity (flexibility) of the model is determined by the number of adjustable parameters, and not by the number of variables. Moreover, adding unobserved variables to the model can reduce the number of adjustable parameters to make the model simpler. This is indeed the case with M_R—it allows a smaller range of possible sensory experience than M_{NR} does. The relative-divergence format of epistemic evaluation therefore favors M_R over M_{NR} if the two models generate theories that fit the experience equally well or nearly equally well. The conversion of the argument from the selection of a theory to the selection of a model strengthens the argument because we need not worry about prior probabilities in selecting a model by the relative-divergence format.

Returning to the BIV model, there is now an appealing argument against it. To be more precise, it is an argument against the generic BIV content model $M_{BIV/CON\text{-}G}$, which is much more flexible than the natural world model M_{NW}. For example, $M_{BIV/CON\text{-}G}$ accommodates a body of jumbled sensory experiences with no discernible structure, while M_{NW} does not. Of course, the specific BIV content model $M_{BIV/CON\text{-}S}$ is only as flexible as the natural world model M_{NW}, but this is a submodel of $M_{BIV/CON\text{-}G}$ selected on the basis of the actual body of sensory experience. Using the specific model $M_{BIV/CON\text{-}S}$ (instead of the general model $M_{BIV/CON\text{-}G}$) to compare with the natural world model M_{NW} looks like the maneuver discussed in the previous section, viz. to adjust parameters of a complex model to accommodate the evidence, and then declare that these parameters are non-adjustable to make the model simple. But whether this procedure is legitimate or not depends on the role of accommodation in the process of selecting a model and generating a theory. It is necessary now to investigate the issue of prediction vs. accommodation in the evaluation of models and theories.

7.4 Prediction vs. Accommodation

It is widely thought that successful prediction is more valuable than successful accommodation in some way. We are impressed—rightly it seems—when a theory predicts some observation we would not otherwise expect, and then the prediction proves correct. We are not so impressed when the same theory is formulated afterword to accommodate the observation that is already obtained. For example, we take notice if some theory correctly predicts the exact sequence DS: [G]GBGBBBGGBG in advance, but we may not pay much attention to the same theory if it is only formulated afterward to accommodate the observed sequence. In general we trust a theory that has a track record of successful predictions much more than we do a theory formulated by accommodation. That is the view of *predictivism*, and the challenge for predictivism is to explain the reason for the difference.

Despite strong intuition in its favor, predictivism is not an uncontroversial position. The critics think it is wrong to value the same theory differently depending on the way it is formulated. That seems to be a conflation of the context of discovery and the context of justification (Reichenbach 1938). In the context of justification we should evaluate a theory by its own merits, regardless of the way we come up with it. There have been attempts to reconcile the two conflicting perspectives in the form of *weak* predictivism.[3] According to weak predictivism, accommodation is not intrinsically problematic, but we value it less than successful prediction because accommodation is prone to produce a theory with problematic features. This means that a theory free from these problematic features is not to be faulted even if it is formulated by accommodation. The challenge for weak predictivism is to identify the problematic features and explain why accommodation is prone to produce a theory with those features.

From the perspective of the relative-divergence format of epistemic evaluation, if accommodation is problematic, then it is because the resulting theory has (or tends to have) a greater estimated divergence from the truth. The main question of this section is whether accommodation increases (or tends to increase) the estimated divergence of the theory from the truth, in comparison with successful prediction. Since the relative-divergence format evaluates theories indirectly by evaluating probabilistic models that generate them, I answer this question in two different settings, viz. in the selection of a probabilistic model and in the generation of a theory from the selected model.

To begin with the selection of a model, when we formulate a theory, we are implicitly restricting a model. Of course, a theory does not single out a model because the same theory can be generated by many models. For example, the quadratic theory $y = 2 + 3x + x^2$ can be generated not just by the quadratic model $y = a_0 + a_1x + a_2x^2$ but by any polynomial model of degree 3 or higher, e.g. the cubic model $y = a_0 + a_1x + a_2x^2 + a_3x^3$ with the parameter value $a_3 = 0$. More generally, any model of which the quadratic model is a separable component can generate a quadratic theory with an appropriate choice of parameter values. However, a theory eliminates some models. For example, the quadratic theory $y = 2 + 3x + x^2$ eliminates the linear model $y = a_0 + a_1x$ and the constant model $y = a_0$ because they do not generate $y = 2 + 3x + x^2$ from any data. More generally, the theory eliminates any model of which the quadratic model is not a separable component.

The worry about accommodation in model selection is the implicit elimination of simpler models. In formulating a theory that accommodates the data well, we are disposed to favor a complex theory (a theory that only complex models can generate) because simpler models with fewer adjustable parameters are less flexible, which impedes our effort to accommodate the data. This is problematic because, as noted in the previous chapter, complex (flexible) models are more prone to chase random variation to generate a theory with a greater divergence from the truth. The problem of accommodation in model selection is then our tendency to eliminate simpler models in the process of accommodation, and this in turn tends to increase the theory's divergence from the truth.[4] This is why close fit between the theory and the data is suspect when a theory is formulated by an accommodation of the data. The theory that eliminates simpler models may accommodate the data better, but its estimated divergence from the truth is often greater than it appears. Note that this account of why accommodation is problematic in model selection supports the position of weak predictivism. The real culprit is the adoption of a more complex model than is appropriate. Though accommodation disposes us to favor complex models, we can resist the temptation with the help of cross-validation, AIC, etc. as discussed in the previous chapter. What matters in the end is not accommodation itself but the complexity of the model.

Let us now turn to accommodation in the generation of a theory. Once a probabilistic model is selected, the model generates a theory from the data, i.e. the values of the adjustable parameters in the model are determined to best fit the data. The generated theory then assigns a probability to a particular outcome, and we evaluate the assignment by its post hoc inaccuracy. There are, however, two cases to distinguish. First, the model may generate a theory from the data that includes the outcome in question, i.e. the outcome in question is taken into account when the values of the parameter are determined. That is a case of accommodation in the generation of a theory. Second, the model may generate a theory from the data that does not include the outcome in question, i.e. the values of the parameters are determined independently of the outcome in question. That is a case of prediction with regard to the outcome.

The difference between the two cases is usually small when the data size is large, but it is still conceptually important. When the outcome fits the theory well but the theory is formulated in part by the outcome in question, we cannot take the post hoc inaccuracy at its face value. For example, when the outcome is an outlier due to random variation, the values of the parameters are adjusted in light of the outcome, which pushes the theory away from the truth. Of course, that is not always the case, but adjusting the values of the parameters in light of the outcome in question creates a bias, i.e. on average it makes the theory appear closer to the truth than it actually is. There is no such bias in the case of prediction, where the values of the parameters are determined independently of the outcome in question. It is for this reason that we split the data set into the training data and the validation data in cross-validation. If a theory to evaluate is generated by the entire data set, we need to mark down its average post hoc inaccuracy to counterbalance the bias due to accommodation. In short, accommodation creates a bias when the model generates a theory.

I want to add some remarks here on the problem of old evidence (Glymour 1980) because the analysis of the bias due to accommodation in the generation of a theory points to its solution. The problem of old evidence arises when a theory receives support from "old evidence" that is known before the theory is formulated. For example, the general theory of relativity received strong support from the precession of the perihelion of Mercury, which was well known before the theory was proposed. However, Bayesian epistemology has difficulty accounting for the role of old evidence. To use the standard terminology, the hypothesis h is confirmed by the evidence e (in the incremental sense of receiving additional support from e) against the background information b just in case e raises the probability of h modulo b, or $P(h \mid b) < P(h \mid e \wedge b)$. The trouble is that when e is old evidence, it is already part of the background information b, so that e in $e \wedge b$ is redundant. It follows immediately that $P(h \mid b) = P(h \mid e \wedge b)$. There is therefore no confirmation of the hypothesis by old evidence, i.e. no additional support from old evidence.

This is actually sensible in some cases. If the hypothesis receives additional support from old evidence, the scientist can cite the same old evidence over and over again to boost the probability of the hypothesis higher and higher indefinitely. However, the dismissal of old evidence is problematic when the theory explains the otherwise puzzling observation made in the past. The theory should receive additional support from the old observation in such cases. Some Bayesians suggest that we loosen the standard Bayesian assumption of logical omniscience to solve the problem (Garber 1983), i.e. old evidence counts if its relation to the theory, which would be known to a logically omniscient reasoner, was unknown to the scientist when the theory is formulated. Such an account may solve "the historical problem of old evidence", which is to explain a change in the degree of confidence on the part of the scientist in the midst of research, but it does not solve "the ahistorical problem of old evidence", which is to explain the value of old evidence in retrospect.[5] With regard to the ahistorical problem of old evidence, Bayesians may pursue the counterfactual account: If the evidence in question had not been known, the discovery of it would have confirmed the theory. Unfortunately, the account is messy and fraught with problems (Glymour 1980).

The analysis of accommodation in the generation of a theory points to a simple solution. It is noted earlier that if the probabilistic model generates a theory from the data that does not include the outcome, then we can take the post hoc inaccuracy at its face value. I called it "prediction", but as is clear from the way cross-validation works, the outcome that is not included in the data need not be collected after the theory is generated. Even if the outcome is old evidence, there is no bias as long as it is not taken into account when the values of the parameters are determined. Take the case of the general theory of relativity. When the theory was formulated, the precession of the perihelion of Mercury was already known, but the values of the parameters in the general theory of relativity, such as the gravitational constant and the speed of light, are determined without consulting the rate of the precession. The old evidence in the case plays the role of the validation data, as distinguished from the training data. So, when the post hoc inaccuracy of the theory turns out to be small with regard to the rate of the precession, we can take the small inaccuracy at its face value. Accommodation in the generation of a theory occurs only when the outcome is not removed from the data when the values of the parameters are determined. This explains why old evidence is as valuable as successful prediction in cases like the precession of the perihelion of Mercury.

7.5 The Natural World and the Virtual World

With the analysis of accommodation in hand, let us return to the issue of the specific BIV content model $M_{BIV/CON\text{-}S}$ that is of the same complexity (flexibility) as the natural world model M_{NW}, but is formulated by accommodation after a body of evidence (sensory experience) is obtained. The question is whether we should mark down $M_{BIV/CON\text{-}S}$ for the reason of accommodation. The answer is no. As shown earlier, accommodation is not problematic in itself in the selection of a probabilistic model. It is true that we are disposed to favor a complex theory (a theory that only complex models can generate) in the process of accommodation. However, accommodation is not a problem if it does not make the model more complex than is appropriate. Since the specific BIV content model $M_{BIV/CON\text{-}S}$ and the natural world model M_{NW} are of the same complexity (flexibility) and fit the evidence equally well, there is no reason for marking down $M_{BIV/CON\text{-}S}$ in comparison with M_{NW}.

In fact it is part of the normal process of model selection to abandon a flexible model we start with in favor of a less flexible model as we obtain more evidence. To use the example of polynomial models, we may begin with the QUAD model $y = a_2x^2 + a_1x + a_0 + \varepsilon$.[6] However, if the best value of the parameter a_2 given the data is zero, or close to zero and dropping it in favor of the LIN model $y = a_1x + a_0 + \varepsilon$ hardly decreases the goodness of fit with the data, then the relative-divergence format recommends that we replace the former by the latter in light of the reduced complexity (flexibility) of the latter. In the same way, the best BIV content model to adopt is determined by the evidence. There is nothing wrong to change a model in light of the evidence unless we make the model overly complex (flexible) in the process, which is not the case when we abandon $M_{\text{BIV/CON-G}}$ in favor of $M_{\text{BIV/CON-S}}$.

The example of starting with QUAD and replacing it by LIN is a good reminder that our model selection evolves with the evidence we acquire, but the example is not analogous to the selection of $M_{\text{BIV/CON-S}}$ in other respects. For example, unlike QUAD, the model $M_{\text{BIV/CON-G}}$ we start with would not have a clear structure with a definite number of parameters. The process of selecting the best model of the BIV content is initially similar to the process of selecting the best *generic* model, only whose submodels have a clear structure with a definite number of parameters. Take the generic model POLY(n) below, of which LIN and QUAD are submodels with $n = 1$ and $n = 2$, respectively.

$$\text{POLY}(n)\text{: } y = \sum\nolimits_{i=0}^{n} a_i x^i + \varepsilon$$

It is straightforward to select a particular polynomial model, such as LIN or QUAD, given the data: We let each model generate a theory from the evidence and compare the estimated divergences of the generated theories from the truth. But how do we evaluate POLY(n) itself in comparison with other generic models? For example, we may want to compare the two generic models, POLY(n) above and TRIG-POLY(n) below.

$$\text{TRIG-POLY}(n)\text{: } y = a_0 + \sum\nolimits_{i=1}^{n} a_i \cos(ix) + \sum\nolimits_{i=1}^{n} b_i \sin(ix) + \varepsilon$$

The relative-divergence format does not seem appropriate for evaluating these generic models because they do not have a definite number of adjustable parameters.

Fortunately, we can still evaluate generic models indirectly by the relative-divergence format as follows. First, out of many submodels that belong to each generic model, we select the best submodel in the usual way, i.e. we let each model (n = 1, 2, ...) generate a theory from the evidence, and estimate their relative divergence from the truth. Once we select the best submodel for each of the candidate generic models, we select the best generic model by comparing their best submodels, i.e. we select the generic model whose best submodel is the best among the best submodels of the candidate generic models. Generic models, such as POLY(n) and TRIG-POLY(n), are boundlessly complex in the sense that its submodel can be as complex (flexible) as the evidence warrants without a limit, but their best submodels selected by the relative-divergence format are usually of moderate complexity, and that allows us to evaluate the generic models indirectly by the relative-divergence format.

The generic version of the BIV content model is also boundlessly complex. Even its direct submodels may not have a definite number of parameters by which to determine their complexity. But that is not an obstacle to applying the relative-divergence format to them. To select the best direct submodel, we identify the best subsubmodel of each candidate submodel, and choose the submodel whose best subsubmodel is the best among them. We repeat the cycle till we reach the bottom layer of models, each of which has a definite number of parameters. The same is true of the natural world model M_{NW}, though it is not as flexible as the model $M_{BIV/CON\text{-}G}$, of which $M_{BIV/CON\text{-}S}$ is a submodel. When the model $M_{BIV/CON\text{-}S}$ replaces the model $M_{BIV/CON\text{-}G}$ to compete with the model M_{NW}, the BIV content model and the natural world model are on a par whose best theories have the same estimated divergence from the truth.

The point that the best version of $M_{BIV/CON}$ is on a par with M_{NW} also applies to other scenarios of Cartesian skepticism. I have been using the BIV hypothesis to illustrate the challenge of Cartesian skepticism, but there are many other scenarios the skeptic can use for the same purpose. The best known among them is the evil demon hypothesis, which I take to be a probabilistic model for the same reason I take the BIV hypothesis to be a probabilistic model. Just as the BIV model consists of $M_{BIV/CON}$ that describes the content of our sensory experience and $M_{BIV/PRO}$ that describes the causal process that produces our sensory experience, the evil demon model also consists of $M_{DEM/CON}$ that describes the content of our sensory experience and $M_{DEM/PRO}$ that describes the causal process that produces our sensory experience. By the same reasoning as above, the best version of $M_{DEM/CON}$ is on a par with M_{NW}. In fact, the content parts of any alternative scenarios of Cartesian skepticism, such as $M_{BIV/CON}$ and $M_{DEM/CON}$, are not just on a par with M_{NW}, they are identical to each other. I will call the shared content part of these models "the virtual world model M_{VW}". The main point made in this section is the parity between the natural world model M_{NW} and the virtual world model M_{VW}. What differentiates different skeptical scenarios is the causal process they propose, e.g. the BIV model combines M_{VW} with the process model $M_{BIV/PRO}$ that a supercomputer is programed to generate sensory experience in BIVs, while the evil demon model combines M_{VW} with the process model $M_{DEM/PRO}$ that an evil demon produces sensory experience in bodiless souls. As we will see in the next two sections, it is the causal part of the model that leads to the demise of Cartesian skepticism.

7.6 Ancestor Simulation

The causal process responsible for our sensory experience is a vital part of any skeptical scenario. It provides an alternative explanation of how we come to have our sensory experience in the absence of normal interactions with the natural environment. However, the skeptic typically provides only a broad outline of the process. For example, the BIV hypothesis states that a supercomputer is programed to generate sensory experience in BIVs, but no details are provided on the type of supercomputer (hardware), the programing language (software) or how the computer is connected to the BIVs (interface). The evil demon hypothesis states that an evil demon produces sensory experience in bodiless souls, but no details are provided on the nature of the evil demon, the nature of the bodiless souls, and the way the evil demon produces sensory experience in the bodiless souls.

This is in sharp contrast with the natural world hypothesis. Scientific investigations have uncovered important details of the causal process responsible for our sensory experience, and as we saw in Chapter 3 there is no problem of epistemic circularity in the empirical (sense perceptual) investigation of the causal process of sense perception. Of course, no empirical investigation can differentiate empirically equivalent hypotheses, but if we set aside skeptical scenarios, we can learn a lot about the causal process of sense perception because the causal process is part of the natural world. The BIVs can also investigate the causal process of sense perception, but it is not the process responsible for the production of their sensory experience. It is only a process in the virtual world to which the BIVs do not belong. The causal process responsible for their own sensory experience is not amenable to empirical investigation.[7] To put the contrast in terms of the model structure, the skeptic must combine the virtual world model M_{VW} with a process model, such as $M_{\mathrm{BIV/PRO}}$ or $M_{DEM/\mathrm{PRO}}$, but there is no need to combine the natural world model M_{NW} with a process model because the causal process responsible for our sensory experience is part of the natural world.

Here is then the state of the dispute in Cartesian skepticism. The natural world model M_{NW} and the virtual world model M_{VW} are on a par, and there is no reason to favor one over the other. However, any skeptical scenario must combine M_{VW} with a model of the causal process responsible for our sensory experience. This gives rise to a problem, viz. we are unable to fill out the details of the process model since the causal process takes place outside the virtual world. One possibility is that the skeptic tells an arbitrary story, e.g. the BIVs receive sensory input by wireless transmission from a quantum computer running Grover's algorithm. I will take up this possibility in the next section, but there is also an interesting suggestion of filling out the process part of the BIV scenario based on empirical evidence. I want to address this possibility in this section.

Bostrom (2003) considers a scenario in which "posthumans" with an enormous amount of computational power simulate lives of their ancestors on computers.[8] Since it is an ancestor simulation, the virtual world and the world in which posthumans live are similar except that the latter is technologically more advanced to run realistic computer simulations of the world their ancestors went through. So, if what we take to be the natural world is a virtual world created by posthumans, we can guess what the posthuman technology is like by extrapolating from the kind of technology available in the virtual world. It is only a rough estimate because there can be radical innovations unforeseen at the time of the simulated ancestors, but it is not a completely arbitrary story. Let us call it "the simulation model M_{SIM}".

The simulation model M_{SIM} has a serious flaw as an alternative scenario for Cartesian skepticism, viz. given the non-perfect reliability of computer simulations *in* the virtual world, we have good reason to think that the ancestor simulations are not perfectly reliable either. We are familiar with programming errors, hardware failure, power outage, etc. that make simulations unreliable. Future devices used in the ancestor simulations may be more reliable, but we cannot eliminate the possibility of errors in simulation. As a result, we always have an option in the simulation model M_{SIM} to blame anomalies on glitches in the causal process outside the virtual world, instead of revising the model of the virtual world (the world the simulation is intended to produce) or adjusting its parameter values. Since that is not an option in the natural world model M_{NW}, which has no additional process component to supplement it, the two models, M_{SIM} and M_{NW}, do not generate the same theory from the same empirical evidence. The additional uncertainty in the causal process makes the M_{SIM}-generated theory predict future observations with less certainty than does the M_{NW}-generated theory. This means that the two models are amenable to comparative evaluation based on empirical evidence. The simulation model M_{SIM} is therefore not a Cartesian skeptical scenario in the traditional sense.

Of course, that the simulation model M_{SIM} is not a Cartesian skeptical scenario in the traditional sense does not mean that it should be rejected. In fact the additional complexity allows the M_{SIM}-generated theory to fit the evidence better than the M_{NW}-generated theory does. However, as we learned in Chapter 6, the additional complexity increases the generated theory's estimated divergence from the truth, which in this case far outweighs a small advantage in the goodness of fit with the evidence, because the model of the causal process $M_{SIM/PRO}$ must posit so many adjustable parameters about hardware, software and interface. If they like, the advocates of the simulation model M_{SIM} can make an empirical case for the model, but there is no good reason at this point to take it seriously by the relative-divergence format of epistemic evaluation.[9]

It may be suggested that though the $M_{SIM/PRO}$ component makes M_{SIM} more complex, the model M_{SIM} as a whole can offset the additional complexity by streamlining its M_{VW} component. In order to deceive BIVs into thinking that they live in a natural world, it is not necessary for the computer program to simulate an entire world. Like the refrigerator light that only turns on when we open the door, the computer program may only simulate a small portion of the world that draws the BIV's attention. That should be sufficient for deception, and make the computer program much simpler than the natural world. However, this suggestion is of no help for the simulation advocates. What counts in the relative-divergence format is not the complexity of the world from which we obtain evidence, but the complexity of the model that generates a theory from the evidence. Making the program lean to produce the same evidence with limited resources does not simplify the M_{VW} model. The M_{SIM} model that consists of M_{VW} and $M_{SIM/PRO}$ is therefore more complex than the M_{NW} model.

7.7 Solution to Cartesian Skepticism

The hypothesis of ancestor simulation allows us to fill out the details of the causal process responsible for our sensory experience in a way that is not completely arbitrary, but it makes the model more complex than the natural world model and the theory it generates is not empirically equivalent to the theory generated by the natural world model. The more traditional form of Cartesian challenge tells an arbitrary story. The skeptic could defend the arbitrariness of the story by arguing that the way we come up with a hypothesis in the context of discovery has no bearing on its epistemic merits—we should evaluate any hypothesis solely by what it proposes, regardless of the way the proposal is conceived. This is a legitimate point with regard to *models*, i.e. the way we come up with a model has no bearing on the model-generated theory's estimated divergence from the truth. However, the point does not apply to *theories*. If there is no model that generates an arbitrarily formulated theory from the actual body of evidence, then it is not a candidate of evaluation by the relative-divergence format, which takes into account the complexity of the model that generates the theory.

So, the Cartesian skeptic must come up with a model that generates a suitable theory from the actual body of evidence, where suitable means that the theory's estimated divergence from the truth is comparable to that of the theory generated by the natural world model. This is not easy because the estimated divergence is affected by the complexity of the model. The difficulty is not the formulation of a theory that fits the data (our sensory experience) as well as the theory generated by the natural world model does. The virtual world model M_{VW}, which is the content part of the BIV model, generates a theory that is empirically equivalent to the theory generated by the natural world model M_{NW}. The two models M_{VW} and M_{NW} are equally complex as well. However, the BIV model also comprises the process part that describes how our sensory experience is produced, and the process model $M_{BIV/PRO}$ added to M_{VW} makes the BIV model as a whole more complex than is the natural world model M_{NW}. The problem for the skeptic is essentially the same as we encountered in the ancestor simulation. Whether the details are filled out on the basis of empirical evidence or filled out arbitrarily, the process part of the model makes the BIV model more complex than the natural world model. The additional complexity allows the M_{BIV}-generated theory to fit the evidence slightly better, but the increase in the estimated divergence from the truth due to complexity outweighs a small advantage in the goodness of fit with the evidence.

There are still some options open to the skeptic who chooses to tell an arbitrary story of the causal process. For example, the skeptic may propose a deterministic process model with no random variation. Recall that the relative-divergence format of epistemic evaluation discounts probabilistic models for complexity because complex (flexible) models are more prone to chase random variation. If the model is deterministic with no random variation, then there is no reason to discount the model for complexity. Of course, the skeptic cannot arbitrarily assign the value one or zero to the probabilistic parameters to make it deterministic because we cannot ignore the actual data when we determine the values of the parameters. We must assign to adjustable parameters those values that best fit the data. Instead, the suggestion is to include only non-probabilistic parameters in the process model. For example, LIN: $y = a_1x + a_0$ is allowed, but the probabilistic LIN: $y = a_1x + a_0 + \varepsilon$ with an error term is not. In the absence of an error term, two non-identical data points $<x_1, y_1>$ and $<x_2, y_2>$ determine the values of a_1 and a_0 in LIN: $y = a_1x + a_0$ completely and conclusively. If some other data point inconsistent with these values is obtained, the model is rejected instead of adjusting the parameter values to make the graph fit the data as close as possible overall. The point is that the deterministic model leaves no room for random variation.

Unfortunately for the skeptic, the BIV model with a deterministic process model is still a probabilistic model overall due to the probabilistic virtual world model M_{VW}. Since the theory generated by the entire BIV model predicts the outcome (sensory experience to come) probabilistically, we can adjust the non-probabilistic parameters in the process model (as well as any parameters in the virtual world model) to fit the data better. So, non-probabilistic parameters in the process model still count for the complexity of the BIV model. The same issue therefore remains: The BIV model is more complex than the natural world model and the complexity makes it inferior to the natural world model.

It may be suggested that the BIV model will be no more complex than the natural world model if we eliminate all adjustable parameters from the process part of the model. That is not possible. First, there are no exogenous parameters in the context of Cartesian skepticism.[10] All parameters in the BIV model are adjustable. So, to eliminate all adjustable parameters from the process model, the skeptic must eliminate all parameters from it. However, our sensory experience has various quantities and degrees—the size, the weight, the degree of brightness, the degree of warmth, etc.—and these quantities are related to each other in various ways. There are therefore quantitative variables to relate in the model, so that the model of the causal process needs some parameters to relate the values of these variables.

Another option the skeptic may pursue is to incorporate perfect reliability into the BIV model. Consider the world the computer program is *intended* to generate in our minds. We may call it "the intended world" to distinguish it from the virtual world we construct from our sensory experience. The causal process that produces our sensory experience is *perfectly reliable* if the virtual world is exactly the intended world. The skeptic may then make the following proposal. It is part of the BIV model that the causal process is perfectly reliable, so that only the content part M_{VW} of the model is responsive to the evidence, while the details of the process part are irrelevant because whatever parameters there are in the process model, they cancel out each other to produce the virtual world in our mind that is exactly the intended world.[11] In addition to being a natural reading of the BIV hypothesis in the literature where it is customary to (tacitly) assume perfect reliability, this version of the BIV model has two features desirable for the skeptic. First, if only the content part M_{VW} of the model is responsive to the empirical findings, the BIV model as a whole becomes empirically equivalent to the natural world model M_{NW}. Second, if the process part of the model is not responsive to any empirical input, then the model is only as complex (flexible) as the content part M_{VW} of the model, and thus only as complex (flexible) as the natural world model M_{NW}. To combine these points, the theories that the two models generate from the actual evidence have the same estimated relative divergence from the truth. There is therefore no reason to prefer the natural world model to this variant of the BIV model.

The question is whether this is a legitimate model amenable to epistemic evaluation by the relative-divergence format. The answer is no. If the causal process is part of a legitimate model, then we cannot simply declare that its parameters cancel out each other to make the causal process perfectly reliable. The values of any parameters in the model must be determined by the data since no parameters are exogenous in the context of Cartesian skepticism. It may be suggested in response that the provision of perfectly reliability is introduced for the reason of simplicity, which is an important part of epistemic evaluation by the relative-divergence format. This response conflates a model and a theory generated by the model. If the values of the parameters, as determined from the data, happen to make the causal process perfectly reliable, then the process model is still part of the BIV model, but then the complexity of the BIV model is not reduced even if the causal process is perfectly reliable. The suggestion here, on the other hand, is to *stipulate* that the causal process is perfectly reliable to make the process part of the model unresponsive to the empirical input, which amounts to dropping the process part of the BIV model. The stipulation reduces the complexity of the model, but that is only because the model no longer contains the process model that is essential to the rejection of the representational parsing of the sensory input. In order to construct an alternative to the natural world hypothesis, the skeptic must posit the causal process outside the represented world.

It is helpful here to recall the example in Chapter 6, where the model LIN2 fails to generate any theory because no value of the independent variable x_2 is available.

LIN2: $y = a_2x_2 + a_1x_1 + a_0$

It was mentioned there that by stipulating $a_2 = 0$ we can make the value of x_2 irrelevant, so that the model can generate a theory without the value of x_2. However, the stipulation amounts to abandoning LIN2 in favor of the simpler model LIN because a_2x_2 plays no role in the model.

LIN: $y = a_1x_1 + a_0$

In the same way, stipulating that the causal process is perfectly reliable eliminates the process part of the model because only the content part of the model responds to the empirical input. The process part plays no role in the model.

In the absence of a process model that specifies the shape of a theory, we are only left with an arbitrary *theory* of the causal process that no model generates from the data. Any arbitrary theory of the causal process—any hardware, any software and any interface—is compatible with any empirical input. There is actually no need to tell a version of the BIV story. Any story—involving an evil demon, a mind-controlling guru, a malevolent cat, etc.—is just as good.[12] The skeptic tells a completely arbitrary story, and just adds that the causal process is perfectly reliable to shield the story from empirical scrutiny. It may be suggested that the skeptic is offering a *generic* deceiver model, viz. an entity of some kind (a supercomputer, an evil demon, etc.) produces the sensory experience of a world in epistemic subjects of some kind (brains in vats, bodiless souls, etc.) by a perfectly reliable method of some kind. But that is not a generic *model*. If it is a generic model, we can combine it with the empirical data to select a submodel, a subsubmodel, etc. to eventually select a theory, but there is no way of selecting any particular theory here. The provision of perfect reliability blocks any attempt to select a theory of the causal process. This is in contrast with the natural world model M_{NW} that generates a detailed theory of the causal process in response to the empirical evidence. The skeptic only proposes an arbitrary theory that no model generates from the data. The deceiver model with the provision of perfect reliability is therefore not a legitimate model amenable to epistemic evaluation by the relative-divergence format.

To conclude, the BIV model fails because of its process component. Since the content component of the model is on a par with the natural world model, and the natural world model does not need an additional process component, the process component of the BIV model makes the entire BIV model more complex than is the natural world model. It is this added complexity (flexibility) that makes the BIV model inferior to the natural world model by the relative-divergence format of epistemic evaluation. The skeptic may try to make the model no more complex than the natural world model by the provision of a perfectly reliable process, but perfect reliability insulates the proposal from any empirical input, and we are left with no model of the causal process to generate a theory from the data. The Cartesian skeptic's proposal is therefore either a model that is inferior to the natural world model, or a model devoid of an account of how our sensory experience is produced. Cartesian skepticism is thereby defeated, and our belief in the natural world is vindicated.

Notes

1 The models in the example are probabilistic. The adoption of a probabilistic model does not restrict the discussion because a probabilistic model can generate a deterministic theory with an extremal probability, zero or one. Note also that even if we suspect the underlying relation is deterministic and we physically control confounding factors by an experimental design, there always remain uncertainties about the accuracy of measurement to make the observation of the expected outcome probabilistic.

2 BonJour (1985, Sec. 8.4) calls the two versions "simple demon hypotheses" and "elaborated demon hypotheses". The term "simple" is not suitable for my purpose because simple hypotheses (with less restrictions) in BonJour's sense are generated by less simple (more flexible) models in my sense. Huemer (2016) calls the two versions "the broad interpretation" and "the narrow interpretation". I do not follow his terminology either because Huemer distinguishes them not only by their different ranges of virtual worlds, but also by their different levels of specificity with regard to the causal process that produces the virtual world. I set aside the issue of the causal process for now and focus on the contents of the sensory input in this section.

3 See, for example, Lange (2001), Hitchcock and Sober (2004) and Harker (2008).

4 A similar analysis of accommodation is proposed by Hitchcock and Sober (2004). They understand model selection in terms of predictive accuracy, but I do not share their instrumentalist orientation. I prefer to frame the issue in terms of the divergence from the truth.

5 The distinction between the historical and the ahistorical problems of old evidence is introduced by Garber (1983). Garber is more concerned with the historical problem, but I consider it to be of secondary importance in comparison with the ahistorical problem of old evidence.

6 The models considered here are all probabilistic models with the error term ε.

7 Veber (2015) points out that many reasons the advocates of the natural world hypothesis cite against the BIV hypothesis have symmetrical counterparts that the advocates of the BIV hypothesis can cite against the natural world hypothesis. One asymmetry missing in Veber's comparison, however, is the epistemic access to the causal process responsible for sensory experience. Brains in skull have an epistemic access to the causal process responsible for their sensory experience, but BIVs have no epistemic access to the causal process responsible for their sensory experience.

8 Bostrom's scenario does not involve brains in vats. He assumes that conscious experiences supervene on functional (computational) properties, so that "a computer running a suitable program would in fact be conscious" (2003, p. 244). I grant this for the argument's sake because it does not affect the main points of the present discussion.

9 Bostrom (2003) argues that an ancestor simulation is probable, but our concern here is the model-generated theory's estimated divergence from the truth. Of course, it is conceivable that the simulation model M_{SIM} turns out to be superior to the model M_{NW} overall despite its additional complexity. Imagine, for example, we are aware that ancestor simulations are already running in some labs, and that some inherent limitations in computer simulation produce certain anomalies in the virtual world. If we find the same anomalies in our own world, we need to take the model M_{SIM} seriously as an explanation of the anomalies.

10 See Section 2 in this chapter for the distinction between endogenous and exogenous parameters.

11 There is no need to introduce a separate model for the intended world. Under the condition of perfect reliability, the model M_{VW} of the virtual world is also a model of the intended world.

12 How can a malevolent cat produce the sensory experience of a virtual world in our mind with perfect reliability? Remember that the causal process that produces our sensory experience may be governed by some laws of nature that are radically different from those by which the virtual world is governed. The skeptic can tell an arbitrary story of the causal process in total disregard of the laws of nature familiar to us.

References

Akaike, H. (1974). A new look at the statistical model identification. *IEEE Transactions on Automatic Control*, *19*, 716–723.

Alston, W. (1986). Internalism and externalism in epistemology. *Philosophical Topics*, *14*, 179–221.

Alston, W. (1988). An internalist externalism. *Synthese*, *74*(3), 265–283.

Alston, W. (1989). *Epistemic justification*. Ithaca, NY: Cornell University Press.

Alston, W. (1993). *The reliability of sense perception*. Ithaca, NY: Cornell University Press.

Alston, W. (2005). *Beyond justification: Dimensions of epistemic evaluation*. Ithaca, NY: Cocrnell University Press.

Ariely, D. (2008). *Predictably irrational: The hidden fores that shape our decisions*. New York: HarperCollins.

Arlot, S., & Celisse, A. (2010). A survey of cross validation procedures for model selection. *Statistics Surveys*, *4*, 40–79.

Armstrong, D. (1978). *A theory of universals: Universals and scientific realism volume II*. Cambridge: Cambidge University Press.

Armstrong, D. (1989). *Universals: An opinionated introduction*. Boulder, CO: Westview Press.

Armstrong, D. (1997). *A world of states of affairs*. Cambridge: Cambridge University Press.

Atkinson, D. (2012). Confirmation and justification. A commentary on Shogenji's measure. *Synthese*, *184*(1), 49–61.

Bar-Hillel, Y., & Carnap, R. (1953). Semantic information. *The British Journal for the Philosophy of Science*, *4*, 147–157.

Bar-On, D., & Simmons, K. (2007). The use of force against deflationism: Assertion and truth. In D. Greimann & G. Siegwart (Eds.), *Truth and speech acts: Studies in the philosophy of language* (pp. 61–89). London: Routledge.

Berkeley, G. (1948). An essay towards a new theory of vision. In A. A. Luce & T. E. Jessop (Eds.), *The works of George Berkeley, bishop of cloyne Vol. 1* (pp. 171–239). London: Thomas Nelson and Sons.

Berker, S. (2013a). Epistemic teleology and the separateness of propositions. *The Philosophical Review*, *122*(3), 337–393.

Berker, S. (2013b). The rejection of epistemic consequentialism. *Philosophical Issues*, *23*(1), 363–387.

Bermúdez, J. L. (1998). *The paradox of self-consciousness*. Cambridge, MA: MIT Press.

Bermúdez, J. L. (2001). Nonconceputal self-consciousness and cognitive science. *Synthese*, *129*, 129–149.

Bermúdez, J. L. (2002). The sources of self-consciousness. *Proceedings of the Aristotelian Society*, *102*, 87–107.

Bernardo, J. (1970). Expected information as expected utility. *Annals of Statistics*, *7*, 686–690.

Block, N. (1986). Advertisement for a semantics for psychology. *Midwest Studies in Philosophy*, *10*, 615–678.

BonJour, L. (1985). *The structure of empirical knowledge*. Cambridge, MA: Harvard University Press.

Bostrom, N. (2003). Are we living in a computer simulation? *Philosophical Quarterly*, *53*, 243–255.

Bovens, L., & Hartmann, S. (2002). *Bayesian epistemology*. Oxford: Oxford University.

Bowers, K., Regehr, G., Balthazard, C., & Parker, K. (1990). Intuition in the context of discovery. *Cognitive Psychology*, *22*, 72–110.

Brandom, R. (1998). Insights and blindspots of reliabilism. *Monist*, *81*, 371–392.

Brown, H. (1993). A theory-laden observation can test the theory. *The British Journal for the Philosophy of Science*, *44*(3), 555–559.

Brown, H. (1994). Circular justifications. *PSA*, *1*, 406–414.

Burge, T. (1979). Individualism and the mental. *Midwest Studies in Philosophy*, *4*, 73–121.

Burnham, K. P., & Anderson, D. R. (2002). *Model selection and multimodel inference: A practical information-theoretic approach*. New York: Springer.

Byrne, A. (2005). Introspection. *Philosophical Topics*, *33*(1), 79–104.

Byrne, A. (2012). Knowing what I see. In S. Declan & D. Stoljar (Eds.), *Introspection and consciousness* (pp. 183–210). Oxford: Oxford University Press.

Cappelen, H. (2012). *Philosophy witout intuitions*. Oxford: Oxford University Press.

Cartwright, R. (1987). *Philosophical essays*. Cambridge MA: MIT Press.

Cevolani, G. (2017). Fallibilism, verisimilitude, and the preface paradox. *Erkenntnis*, *82*(1), 169–183.

Cevolani, G., & Schurz, G. (2017). Probability, approximate truth, and truthlikeness: More ways out of the preface paradox. *Australasian Journal of Philosophy*, *95*(2), 209–225.

Chalmers, D. (2003). The content and epistemology of phenomenal belief. In Q. Smith & A. Jokić (Eds.), *Consciousness: New philosophical perspectives* (pp. 220–272). Oxford: Oxford University Press.

Chalmers, D. (2005). The matrix as metaphysics. In C. Grau (Ed.), *Philosophers explore the matrix* (pp. 132–176). Oxford: Oxford University Press.

Clifford, W. K. (1999). The Ethics of belief. In T. Madigan (Ed.), *The ethics of belief and other essays*. Amherst NY: Prometheus Books.

Climenhaga, N. (forthcoming). Intuitions are used as evidence in philosophy. *Mind*. https://doi.org/10.1093/mind/fzw032.

Coady, C. A. (1992). *Testimony: A philosophical study*. Oxford: Oxford University Press.

Craig, E. (1990). *Knowledge and the state of nature: An essay in conceptual synthesis*. Oxford: Oxford University Press.

Crupi, V., Fitelson, B., & Tentori, K. (2008). Probability, confirmation, and the conjunction fallacy. *Thinking & Reasoning*, *14*(2), 182–199.

Crupi, V., Tentori, K., & Gonzalez, M. (2007). On Bayesian measures of evidential support: Theoretical and empirical issues. *Philosophy of Science*, *74*(2), 229–252.

Cummins, R. (1998). Reflections on reflective equilibrium. In M. R. DePaul & W. Ramsey (Eds.), *Rethinking intuition* (pp. 113–127). Lanham, MD: Rowman & Littlefield.

Davidson, D. (1973). Radical interpretation. *Dialectica*, *27*, 313–328.

Davies, M. (1992). Perceputal content and local supervenience. *Proceedings of the Aristotelian Society*, *92*, 21–45.

de Finetti, B. (1937). La prevision: Ses lois logiques, se sources subjectives. *Annales de l'Institut Henri Poincare*, *7*, 1–68.

Dembroff, R. (2016). What is sexual orientation. *Philosophers' Imprint*, *16*(3), 1–27.

Dennett, D. (1981). Where am I. In *Brainstorms: Philosophical essays on mind and pscyology* (pp. 310–323). Cambridge, MA: MIT Press.

DeRose, K. (1995). Solving the skeptical problem. *The Philosophical Review*, *104*, 1–52.

Descartes, R. (1984). Meditations on first philosophy. In J. Cottingham (Trans.), *The philosophical writings of Descartes, Vol. II* (pp. 12–62). Cambridge: Cambridge University Press.

Descartes, R. (1985a). Rules for the direction of the mind. In D. Murdoch (Trans.), *The philosophical writings of Descartes, Vol. I* (pp. 9–78). Cambridge: Cambridge University Press.

Descartes, R. (1985b). Discourse on the method. In R. Stoothoff (Trans.), *The philosophical writings of Descartes, Vol. I* (pp. 111–151). Cambridge: Cambridge University Press.

Deutsch, M. (2015). *The myth of the intuitive: Experimental philosophy and philosophical method*. Cambridge, MA: MIT Press.

Douven, I. (2005). Evidence, explanation, and the empirical status of scientific realism. *Erkenntnis*, *63*, 253–291.

Douven, I. (2008). Underdetermination. In S. Psillos & M. Curd (Eds.), *The Routledge companion to philosophy of science* (pp. 292–301). London: Routledge.

Dretske, F. (1971). Conclusive reasons. *Australasian Journal of Philosophy*, *49*(1), 1–22.

Dreyfus, H. (1972). *What computers can't do*. Cambridge, MA: MIT Press.

Emery, N. (2015). Chance, possibility, and explanation. *The British Journal for the Philosophy of Science*, *66*(1), 95–120.

Evans, G. (1982). *The varieties of reference* (J. McDowell, Ed.). Oxford: Oxford University Press.

Fantl, J., & McGrath, M. (2002). Evidence, pragmatics, and justification. *The Philosophical Review*, *111*(1), 67–94.

Fantl, J., & McGrath, M. (2007). On pragmatic encroachment in epistemology. *Philosophy and Phenomenological Research*, *75*(3), 558–589.

Fantl, J., & McGrath, M. (2009). *Knowledge in an uncertain world*. Oxford: Oxford University Press.

Field, H. (1977). Logic, meaning and conceptual role. *The Journal of Philosophy*, *69*, 379–409.

Field, H. (1994). Deflationist views of meaning and content. *Mind*, *103*, 249–285.

Firth, R. (1998). Epistemic merit, intrinsic and instrumental. In J. Troyer (Ed.), *In defense of radical empiriccism: Essays and lectures by Frederick Firth* (pp. 259–271). Lanham, MD: Rowman & Littlefield.

Fitelson, B. (1999). The plurality of Bayesian measures of confirmation and the problem of measure sensitivity. *Philosophy of Science*, *66*, S362–S378.

Fitelson, B. (2002). Symmetries and asymmetries in evidential support. *Philosophical Studies*, *107*(2), 129–142.

Foley, R. (1993). *Working without a net*. Oxford: Oxford University Press.

Foley, R. (2009). Beliefs, degrees of belief, and the Lockean thesis. In F. Huber & C. Schmidt-Petri (Eds.), *Degrees of belief* (pp. 37–47). Dordrecht, The Netherlands: Springer.

Foot, P. (1967). The problem of abortion and the doctrine of double effect. *Oxford Review*, *5*, 5–15.

Forster, M., & Sober, E. (1994). How to tell when simpler, more unified, or less ad hoc theories will provide more accurate predictions. *The British Journal for the Philosophy of Science*, *45*(1), 1–35.

Garber, D. (1983). Old evidence and logical omniscience in Bayesian confirmation theory. *Testing Scientific Theories*, *10*, 99–131.

Gärdenfors, P. (2000). *Conceptual spaces: The geometry of thought*. Cambridge, MA: MIT Press.

Gertler, B. (2001). Introspecting phenomenal states. *Philosophy and Phenomenological Research*, *63*(2), 305–328.

Gettier, E. (1963). Is justified true belief knowledge? *Analysis*, *23*, 121–123.

Gibson, J. (1979). *The ecological approach to visual perception*. Boston, MA: Houghton Mifflin.

Glanzberg, M. (2015). Modern correspondence theory of truth. In S. Gross, N. Tebben, & M. Williams (Eds.), *Meaning without representation: Essays on truth, expression, normativity, and naturalism* (pp. 81–102). Oxford: Oxford University Press.

Glymour, C. (1980). *Theory and evidence*. Princeton, NJ: Princeton University Press.

Goldman, A. (1979). What is justified belief. In G. Pappas (Ed.), *Justification and knowledge* (pp. 1–25). Dordrecht, The Netherlands: Reidel.

Goldman, A. (1980). The internalist conception of justification. *Midwest Studies in Philosophy*, *5*, 27–51.

Goldman, A. (1999). *Knowledge in a social world*. Oxford: Oxford University Press.

Goldman, A. (2015). Reliabilism, veritism, and epistemic consequentialism. *Episteme*, *12*(2), 131–143.

Goodman, N. (1955). *Facts, fiction, and forecast*. Cambridge, MA: Harvard University Press.

Gordon, R. (1995). Simulation without introspection or inference from me to you. In M. Davies & T. Stone (Eds.), *Mental simulation* (pp. 53-67). Oxford: Blackwell.

Gordon, R. (1996). 'Radical' simulationism. In P. Carruthers & P. K. Smith (Eds.), *Theories of theories of mind* (pp. 11–21). Cambridge: Cambridge University Press.

Gordon, R. (2007). Ascent routines for propositional attitudes. *Synthese*, *159*, 151–165.

Harker, D. (2008). On the predilections for predictions. *The British Journal for the Philosophy of Science*, *59*, 429–453.

Harman, G. (1989). *Change in view*. Cambridge, MA: MIT Press.

Harman, G. (1990). The intrinsic quality of experience. *Philosophical Perspectives*, *4*, 31–52.

Harman, G. (1999). *Reason, meaning, and mind*. Oxford: Oxford University Press.

Haslanger, S. (2000). Gender and race: (What) are they? (What) do we want them to be? *Noûs*, *34*(1), 31–55.

Haslanger, S. (2005). What are we talking about? The semantics and politics of social kinds. *Hypatia*, *20*(4), 10–26.

Hempel, C. (1960). Inductive inconsistencies. *Synthese*, *12*, 439–469.

Henderson, D., & Greco, J. (Eds.). (2015). *Epistemic evaluation: Purposeful epistemology*. Oxford: Oxford University Press.

Hill, C. (2002). *Thought and world: An Austere portrayal of truth, reference, and semantic correspondence*. Cambridge: Cambridge University Press.

Hill, C. (2014). *Meaning, mind, and knowledge*. Cambridge: Cambridge University Press.

Hill, C. (2016). Deflationism: The best thing since pizza and quite possibly better. *Philosophical Studies*, *173*, 3169–3180.

Hitchcock, C., & Sober, E. (2004). Prediction versus accommodation and the risk of overfitting. *The British Journal for the Philosophy of Science*, *55*(1), 1–34.

Hornsby, J. (1997). Truth: The identity theory. *Proceedings of the Aristotelian Society*, *97*, 1–24.

Horwich, P. (1990). *Truth*. Oxford: Oxford University Press.

Horwich, P. (1998). *Meaning*. Oxford: Oxford University Press.

Huber, F. (2008a). Assessing theories, Bayes style. *Synthese*, *161*, 89–118.

Huber, F. (2008b). Hempel's logic of confirmation. *Philosophical Studies*, *139*(2), 181–189.

Huemer, M. (1997). Probability and coherence justification. *The Southern Journal of Philosophy*, *35*(4), 463–472.

Huemer, M. (2016). Serious theories and skeptical theories: Why you are probably not a brain in a vat. *Philosophical Studies*, *173*(4), 1031–1052.

Hurvich, C. M., & Tsai, C-L. (1989). Regression and time series model selection in small samples. *Biometrika*, *76*(2), 297–307.

James, W. (1979). The sentiment of rationality. *The will to believe and other essays in popular philosophy* (pp. 57–89). Cambridge, MA: Harvard University Press.

Jaynes, E. T. (2003). *Probability theory: The logic of science* (G. L. Bretthorst, Ed.). Cambridge: Cambridge University Press.

Jeffrey, R. (1983). *The logic of decision* (2nd ed.). Chicago: University of Chicago Press.

Jeffreys, H. (1939). *Theory of probability*. Oxford: Oxford University Press.

Jenkins, K. (2016). Amelioration and inclusion: Gender identity and the concept of woman. *Ethics*, *126*(2), 394–421.

Kaplan, M. (1981a). A Bayesian theory of rational acceptance. *The Journal of Philosophy*, *78*(6), 305–330.

Kaplan, M. (1981b). Rational acceptance. *Philosophical Studies*, *40*(2), 129–145.

Kaplan, M. (1991). Epistemology on holiday. *The Journal of Philosophy*, *88*, 132–154.

Kaplan, M. (1996). *Decision theory as philosophy*. Cambridge: Cambridge University Press.

Kitcher, P. (1992). The naturalists return. *The Philosophical Review*, *101*, 53–114.

Knobe, J., & Nichols, S. (2008). *Experimental philosophy*. Oxford: Oxford University Press.

Knobe, J., & Nichols, S. (2013). *Experimental philosophy* (Vol. 2). Oxford: Oxford University Press.

Kohavi, R. (1995). A study of cross-validation and bootstrap for accuracy estimation and model selection. *International Joint Conference on Artificial Intelligence*, *14*(2), 1137–1145.

Kripke, S. (1980). *Naming and necessity*. Cambridge, MA: Harvard University Press.

Kuipers, T. A. (2000). *From instrumentalism to constructive realism: On some relations between confirmation, empirical progress, and truth approximation*. Dordrecht, The Netherlands: Kluwer.

Kyburg, H. (1961). *Probability and the logic of rational belief*. Middletown: Wesleyan University Press.

Lange, M. (2001). The apparent superiority of prediction to accommodation as a side effect: A reply to Maher. *The British Journal for the Philosophy of Science*, *52*(3), 575–588.

Leitgeb, H., & Pettigrew, R. (2010). An objective justification of Bayesianism I: Measuring inaccuracy. *Philosophy of Science*, *77*(2), 201–235.

Levi, I. (1967). *Gambling with truth*. Cambridge, MA: Cambridge University Press.

Levi, I. (1980). *The enterprise of knowledge*. Cambridge, MA: MIT University Press.

Levi, I. (2004). *Mild contraction: Evaluating loss of information due to loss of belief*. Oxford: Oxford University Press.

Lewis, D. (1979). Attitudes de dicto and de se. *The Philosophical Review*, *88*, 513–543.

Lewis, D. (2001). Forget about the 'correspondence theory of truth'. *Analysis*, *61*, 275–280.

Lyons, J. (2016). *Epistemological problems of perception*. In E. N. Zalta (Ed.), *Encyclopedia of philosophy*. Stanford, CA. Retrieved from https://plato.stanford.edu/archives/spr2017/entries/perception-episprob/

Maher, P. (1993). *Betting on theories*. Cambridge: Cambridge University Press.

Makinson, D. C. (1965). The paradox of the preface. *Analysis*, *25*(6), 205–207.

Marr, D. (1982). *Vision: A computational investigation into the human representation and processing of visual information*. New York: W. H. Freeman.

McDowell, J. (1996). *Mind and world*. Cambridge, MA: Harvard University Press.

McGinn, C. (1982). The structure of content. In A. Woodfield (Ed.), *Thought and object* (pp. 207–259). Oxford: Oxford University Press.

McGinn, C. (2002). The truth about truth. In R. Schantz (Ed.), *What is truth?* (pp. 194–204). Berlin: Walter de Gruyter.

McGrew, T., & McGrew, L. (2007). *Internalism and epistemology: The architecture of reason*. New York: Routledge.

Merleau-Ponty, M. (1962). *Phenomenology of perception* (C. Smith, Trans.). London: Routledge.

Miller, D. (1974). Popper's qualitative theory of verisimilitude. *The British Journal for the Philosophy of Science*, *25*(2), 166–177.

Moretti, L., & Shogenji, T. (2017). Skepticism and epistemic closure: Two Bayesian accounts. *International Journal for the Study of Skepticism*, *7*, 1–25.

Murphy, A., & Winkler, R. (1984). Probability forecasting in meteorology. *Journal of the American Statistical Association*, *79*, 489–500.

Nagel, T. (1989). *The view from nowhere*. Oxford: Oxford University Press.

Niiniluoto, I. (1987). *Truthlikeness*. Dordrecht, The Netherlands: Reidel.

Niiniluoto, I. (1999). *Critical scientific realism*. Oxford: Oxford University Press.

Niiniluoto, I. (2013). *Is science progressive?* New York: Springer.

Nozick, R. (1981). *Philosophical explanations*. Cambridge: Cambridge University Press.

Olsson, E. J. (2002). What is the problem of coherence and truth? *The Journal of Philosophy*, *99*(5), 246–272.

Olsson, E. J. (2005). *Against coherence: Tuth, probability, and justification*. Oxford: Oxford University Press.

Peijnenburg, J., & Atkinson, D. (2013). The emergence of justification. *The Philosophical Quarterly*, *63*, 546–564.

Perry, J. (1979). The problem of essential indexical. *Noûs*, *13*, 3–21.

Pettigrew, R. (2016). *Accuracy and the laws of credence*. Oxford: Oxford University Press.

Plantinga, A. (1993). *Warrant and proper function*. Oxford: Oxford University Press.

Pollock, J., & Cruz, J. (1999). *Contemporary theories of knowledge* (2nd ed.). Lanham, MD: Rowman & Littlefield.

Popper, K. (1954). Degree of confirmation. *The British Journal for the Philosophy of Science*, *5*, 143–149.

Popper, K. (1963). *Conjectures and refutations: The growth of scientific knowledge*. London: Routledge.

Popper, K. (1972). *Objective knowledge. An evolutionary approach*. Oxford: Oxford University Press.

Pust, J. (2001). Against explationist skepticism regarding philosophical intuitions. *Philosophical Studies, 106*(3), 227–258,

Pust, J. (2017). *Intuition*. In E. N. Zalta (Ed.), *Stanford encyclopedia of philosophy*. Retrieved from https://plato.stanford.edu/archives/sum2017/entries/intuition/

Putnam, H. (1975). The meaning of 'meaning'. *Minnesota Studies in the Philosophy of Science*, *7*, 215–271.

Ramsey, F. (1931). Truth and probability. In F. Ramsey (Ed.), *Foundations of mathematics and other logical essays* (pp. 156–198). London: Kegan Paul.

Rawls, J. (1971). *A theory of justice*. Cambridge, MA: Harvard University Press.

Rawls, J. (2001). *Justice as fairness: A restatement*. Cambridge, MA: Harvard University Press.

Reichenbach, H. (1938). *Experience and prediction*. Chicago: University of Chicago Press.

Roche, W., & Shogenji, T. (2014). Dwindling confirmation. *Philosophy of Science*, *81*(1), 114–137.

Roche, W., & Shogenji, T. (Forthcoming). Information and inaccuracy. *The British Journal for the Philosophy of Science*.

Schmitt, F. (Ed.). (1994). *Socializing epistemology*. Lanham, MD: Rowman & Littlefield.

Searle, J. (1983). *Intentionality*. Cambridge: Cambridge University Press.

Shannon, C. E., & Weaver, W. (1949). *The mathematical theory of communication*. Urbana, IL: University of Illinois Press.

Shogenji, T. (2000). Self-dependent justification without circularity. *The British Journal for the Philosophy of Science*, *51*, 287–298.

Shogenji, T. (2002). The problem of independence in justification by coherence. In Y. Bouchard (Ed.), *Perspectives on coherentism* (pp. 129–137). Alymer: Éditions du Scribe.

Shogenji, T. (2005). Justification by coherence from scratch. *Philosophical Studies*, *125*(3), 305–325.

Shogenji, T. (2006). A defense of reductionism about testimonial justification of beliefs. *Noûs*, *40*(2), 331–346.

Shogenji, T. (2012). The degree of epistemic justification and the conjunction fallacy. *Synthese*, *184*(1), 29–48.

Shogenji, T. (2013). Reductio, coherence, and the myth of epistemic circularity. In F. Zenker (Ed.), *Bayesian argumentation* (pp. 165–184). New York: Springer.

Sides, A., Osherson, D., Bonini, N., & Viale, R. (2002). On the reality of the conjunction fallacy. *Memory & Cognition, 30,* 191–198.

Siegel, S. (2006). Subject and object in the contents of visual experience. *The Philosophical Review*, *115*(3), 355–388.

Smyth, N. (2017). The function of morality. *Philosophical Studies*, *174*(5), 1127–1144.

Stevenson, L. (1999). First person epistemology. *Philosophy*, *74*, 475–497.

Stich, S. (1983). *From folk psychology to cognitive science.* Cambridge, MA: MIT Press.

Stone, M. (1977). An asymptotic equivalence of choice of model by cross-validation and Akaike's criterion. *Journal of the Royal Statistical Society. Series B (Methodological)*, *36*, 44–47.

Thaler, R., & Sunstein, C. (2008). *Nudge: Improving decisions about health, wealth, and happiness.* New Haven, CT: Yale University Press.

Tichý, P. (1974). On Popper's definitions of verisimilitude. *The British Journal for the Philosophy of Science*, *25*(2), 155–160.

Tversky, A. & Kahneman, D. (1974). Judgment under uncertainty: Heuristics and biases. *Science, 185,* 1124–1131.

Tversky, A., & Kahneman, D. (1983). Extensional veresus intuitive reasoning: The conjunction fallacy in probabilistic judgment. *Psychological Review*, *90*(4), 293–315.

Veber, M. (2015). What's it like to be a BIV? A dialogue. *Journal of the American Philosophical Association*, *1*(4), 734–756.

Watanabe, S. (2009). *Algebraic geometry and statistical learning theory*. Cambridge: Cambridge University Press.

Wheeler, G. (2007). A review of the lottery paradox. In W. Harper & G. Wheeler (Eds.), *Probability and inference: Essays in honour of Henry E. Kyburg, Jr* (pp. 1–31). London: King's College Publications.

Williamson, J. (2010). *In defense of objective Bayesianism.* Oxford: Oxford University Press.

Williamson, T. (2008). *The philosophy of philosophy*. Hoboken, NJ: Wiley-Blackwell.

Winkler, R. (1967). The quantification of judgment: Some methodological suggestions. *Journal of the American Statistical Association*, *62*, 1105–1120.

Winkler, R. (1969). Scoring rules and the evaluation of probability assessors. *Journal of the American Statistical Association*, *64*, 1073–1078.

Winkler, R. (1971). Probabilistic prediction: Some experimental results. *Journal of the Americal Statistical Association*, *66*, 675–685.

Winkler, R. (1994). Evaluating probabilities: Asymmetric scoring rules. *Management Science*, *40*, 1395–1405.

Winkler, R., & Murphy, A. (1968). 'Good' probability assessors. *Journal of Applied Meteorology*, *7*, 751–758.

Wójtowicz, A., & Bigaj, T. (2016). Justification, confirmation, and the problem of mutually exclusive hypotheses. *Poznań Studies in the Philosophy of the Sciences and the Humanities*, *107*, 122–143.

Index